CAN YOU WIN?

CAN YOU WIN?

The Real Odds for Casino Gambling, Sports Betting, and Lotteries

Mike Orkin

W. H. Freeman and Company
New York

Library of Congress Cataloging-in-Publication Data

Orkin, Michael.
    Can you win?: the real odds for casino gambling, sports betting,
  and lotteries / Mike Orkin.
        p.   cm.
    Includes bibliographical references and index.
    ISBN 0-7167-2155-4 (soft)
    1. Probabilities. 2. Gambling. 3. Sports betting.
  4. Lotteries. I. Title.
    GV1302.075 1991
    795'.01 — dc20   90-19284
                                                                          CIP

Printed in the United States of America

Sixth printing, 2001

This book is dedicated to Professor David Blackwell

CONTENTS

Preface   ix

Chapter 1   LIFE IS A GAMBLE   1

Chapter 2   ROULETTE   7

Chapter 3   CRAPS   27

Chapter 4   KENO AND SLOTS   47

Chapter 5   SPORTS BETTING   55

Chapter 6   BLACKJACK   87

Chapter 7   STATE LOTTERIES   97

Chapter 8   HORSE RACING   113

Chapter 9   PRISONER'S DILEMMA   131

Glossary of Gambling Terms   141

Bibliography   145

Figuring the Odds   147

Answers to Figuring the Odds   167

Index   179

PREFACE

This book is about gambling and games. If you're going to gamble, you may as well learn the facts. Which casino games can you beat? What do state lottery odds *really* mean? What's the point of gambling if you'll probably lose? What's the point of doing anything?

In *Can You Win?* I'll explain the rules and odds for popular casino games, sports betting, state lotteries, and horse racing. I'll also discuss the game Prisoner's Dilemma, which sheds light on the problem of how to be nice yet avoid being taken advantage of.

Gambling involves math, but I'll keep it easy. If you like problems and puzzles, you'll want to work the problems in the section at the end of the book called "Figuring the Odds." If you're not interested in a particular section, skip it. If you have no interest in gambling, stop, but don't come crying to me if you lose your life's savings in the state lottery.

ACKNOWLEDGMENTS

I'd like to thank the following people, whose knowledge and suggestions have helped me immensely: David Blackwell, Sonia DiVittorio, Richard Drogin, Leslie Freerks, Dan Gordon, David Hildebrand, Richard Kakigi, Arne Lang, David Levine, Jerry Lyons, George McCabe, "Moth," Mort Olshan, Leon Orkin, Christine Rickoff, Michael Roxborough, Howard Schwartz, and Sylvia Warren.

CAN YOU WIN?

1

· · · · · · ·

LIFE IS A GAMBLE

I hope I break even this week. I need the money.

—VETERAN LAS VEGAS GAMBLER

Gambling is a national pastime. A PBS special on compulsive gambling, aired in June 1990, estimated that $253 billion a year is spent on gambling, $13 million more than all 50 states raise in taxes. Nevada and Atlantic City offer a variety of legalized gambling games. Lotteries generate revenue for numerous states. The stock market provides a respectable betting environment for those who can afford it. At the racetrack people from all walks of life test Damon Runyon's proposition that horseplayers die broke. Illegal sports betting is pervasive. America is dense with bookies.

· ·

If you bet on a horse, that's gambling. If you bet you can make three spades, that's entertainment. If you bet cotton will go up three points, that's business. See the difference?

—BLACKIE SHERROD, DALLAS SPORTSWRITER

In California, you can legally bet your life's savings in the state lottery but be tossed in jail for betting $5 in the office football pool. In this book, I'll tell you the facts about popular gambling games.

To me, gambling is taking a risk. According to *Webster's* dictionary, gambling is "playing a game for money or other stakes." Michael "Roxy" Roxborough, head Las Vegas oddsmaker, defines a gamble as "a wager made at unfavorable odds." Stock speculators may disagree with Roxy's definition, but for most casino games, it's accurate.

Lee the Flea, one of my students by day and professional cupcake baker by night, knows that gambling is best regarded as a form of entertainment. When Lee and his friends enter Nevada, they pretend that money turns into bananas.

Buck and his girlfriend, Dotty, are an intelligent, caring couple. Buck is a retired insurance agent and Dotty a retired psychiatrist. Buck likes to play craps, and Dotty likes to bet on football games. Buck spent three months trying to convince me that he had a winning craps system, even though he knew such a system was mathematically impossible. With a gleam in his eye, Buck told me, "I know it's impossible, but what if it weren't?"

The Lone Wolf is a purist and insists on making a living only from sports betting. He has a photographic memory and will describe for you, in detail, any pro basketball game played in the last decade. The Lone Wolf wears a sports beeper to get instant game updates wherever he goes.

I know an attorney named Moth, who wrote a horse handicapping column for the school paper when we were in college. Moth watched numerous sure winners lose from his grandstand seat at Golden Gate Fields racetrack.

GAMBLING AND PROBABILITY

Probability theory was developed by mathematicians studying gambling games. The same laws of probability that apply to gambling games also apply to genetics, marketing strategies, and military plans.

In 1654, the titled French con man Antoine Gombaud, the Chevalier de Mère, offered even money odds that in four rolls of a die at least one 6 would come up. The Chevalier reasoned that since the chance equals 1/6 that a 6 will come up when a die is rolled, then in four rolls, the chance of getting at least one 6 equals 4/6 = 2/3. The Chevalier's reasoning was faulty (the actual chance of getting at least one 6 in four rolls equals .52), but it didn't cost him. In fact, he made a profit.

When the betting dropped off, the Chevalier modified his game. He reasoned that since the chance equals 1/36 that a double-6 will come up when a pair of dice are rolled once, the chance that at least one double-6 will come up in 24 rolls equals 24/36 = 2/3. This time, the Chevalier paid for his faulty reasoning. He started losing. In desperation, he consulted the mathematician Blaise Pascal.

Pascal's interest in the Chevalier's games spurred a correspondence with the mathematician Fermat. This marked the beginning of modern probability theory.

The Chevalier had a 52% chance of winning the first game but only a 49% chance of winning the second game. This slight change in probabilities marks the difference between riches and rags in repeated play.

THE LAW OF AVERAGES

One of the most important results in probability theory is the law of averages, also known as the law of large numbers. This mathematical law provides an important connection between theoretical probabilities and observed results of random processes. Here is a version of the law of averages:

> In repeated, independent trials of the same experiment, the observed fraction of occurrences of an event eventually approaches its theoretical probability. (Independent trials occur when the outcome of one trial has no effect on the outcome of any other trial.)

Here's an example. If you toss a coin, the theoretical probability that heads will come up equals 1/2. I'll use this notation: $P(\text{heads}) = 1/2$. If you toss the coin repeatedly, even though you don't know whether heads or tails will come up on a particular toss, the actual fraction of heads (and tails) will eventually get close to 1/2. For example, if you toss an ordinary coin 100 times, there's a 95% chance that the fraction of heads will be between .40 and .60 (40–60 heads). Heads will almost certainly come up between 35 and 65 times. Try it.

It follows from the law of averages that if you repeatedly play an "unfavorable" game, even though you're uncertain of the results of an individual play, in the long run you will surely be a loser.

Luck is a group activity. For every lucky winner of an unfavorable gambling game, there are a group of unlucky losers. The chances of winning the California 6 for 49 Lotto jackpot were about 1 in 14 million. It took an average of 14 million players (losers) to produce a winner. The chances of winning the current 6 for 53 jackpot are about 1 in 23 million. On the other hand, even when the chances of winning are extremely small, if enough people play, someone will win. Professors Persi Diaconis and Frederick Mosteller call this phenomenon the law of very large numbers: "With a large enough sample, any outrageous thing is apt to happen."

THE HOUSE EDGE

A casino's long-run average winnings per dollar bet for a particular wager, expressed as a percentage, is called the *house edge*. The law of averages lets you compute the house edge for many games. We will compute the house edge for casino bets. The Chevalier had a 4% edge in his first game. In the second game, the bettors had a 2% edge. In repeated play of the second game, the Chevalier lost 2% of all money bet. Word got around, and the Chevalier got fleeced.

Ironically, if in the second game the Chevalier had allowed 25 rolls of the dice instead of 24 to get a double-6, he would have been a winner. Pascal wouldn't have heard about the problem, and science would have suffered.

In 1952, a New York gambler known as Fat the Butch gave even money odds that in 21 rolls of a pair of dice he would get at least one double-6. Apparently he hadn't heard of the Chevalier. His opponents had a 10% edge. In a series of bets with a gambler known as The Brain, Fat the Butch lost $50,000.

Casinos offer games with a variety of house edges. The house edge for roulette is 5.3%. If you make a roulette bet repeatedly, you will eventually lose at the rate of 5.3 cents per dollar bet. In craps, there are bets with different house edges. The house edge for the pass line bet is 1.4%, while the house edge for the big six bet is 9.1%. In keno, the house edge is more than 25%. The picture is bleak in state lotteries, where the state has a 50% edge. The gambler who thinks he will be a winner in repeated play of these games is like the retailer who sells every item at a loss, hoping to make a profit from the increased volume.

A few gambling games, like blackjack, have winning strategies for the player. But if your criterion for success is a high hourly income, you may end up a failure even when you play a favorable game. I know a professional poker player who says, "A good poker player makes $20 per hour. A *really* good poker player makes $30 per hour."

CAN YOU WIN?

. . . .

Here you are. Furniture and fittings, I'll take four hundred or the nearest offer. . . . You can reckon water, heating and lighting at close on fifty. That'll cost you eight hundred and ninety if you're all that keen. Say the word and I'll have my solicitors draft you out a contract. Otherwise, I've got

the van outside, I can run you to the police station in five minutes, have
you in for trespassing, loitering with intent, daylight robbery, filching,
thieving, and stinking the place out. What do you say?

—HAROLD PINTER, *THE CARETAKER*

Casino games aren't the only games in town. There are games in
which neither player uses a known, fixed strategy the way a casino does.
Real-life game players can be citizens, countries, companies, or com-
puters. In some gamelike situations, players can either be cooperative or
nasty. In such situations, you are not gambling for money, but gambling
that if you behave a certain way, your opponent will behave a certain way.

Interactions involving both cooperation and competition pose this
problem: How can you cooperate yet avoid being taken advantage of?
This is illustrated in a game called Prisoner's Dilemma, which I'll discuss
in Chapter 9.

· · ·

Think of all the games in your daily life, the gambles you take, the
strategies you use. There are many misconceptions about gambling and
strategies. Contrary to popular belief, you *can't* beat most gambling games
in the long run. But gambling can be exciting, even with unfavorable odds.
Wagering a few dollars in a lottery gives you a chance, however small, to
dramatically change your lifestyle. As my mother used to say, "As long as
you're happy."

2
· · · · · · ·

ROULETTE

Feeling as though I were delirious with fever, I moved the whole pile of money to the red—and suddenly came to my senses! For the only time in the course of the whole evening, fear laid its icy finger on me and my arms and legs began to shake. With horror I saw and for an instant fully realized what it would mean to me to lose now! My whole life depended on that stake!" . . .

I elbowed my way into the thick of them and stood close to the croupier; . . . [I]t appeared to me that pure calculation means fairly little and has none of the importance many gamblers attach to it. They sit over bits of paper ruled into columns, note down the coups, count up, compute probabilities, do sums, finally put down their stakes and—lose exactly the same as we poor mortals playing without calculation.

—FYODOR DOSTOYEVSKI, *THE GAMBLER*

Roulette, a symbol of gambling decadence, is played with a large wheel that spins on a shaft mounted next to a betting layout (Figure 1). The wheel is divided into 38 sections, numbered from 1 to 36, 0, and 00. Eighteen of the sections numbered from 1 to 36 are red and eighteen are black. The sections 0 and 00 are green. The sections are colored alternately red and black, with 0 and 00 on opposite sides of the wheel.

The wheel is spun by a casino employee known as the croupier, who then spins a ball along the wheel in the opposite direction. As the wheel and ball slow down, the ball drops to the center of the wheel, finally landing in one of the numbered sections. The section the ball lands in is the winning section.

To make bets in roulette, you buy special roulette chips from the dealer for the denomination you want, subject to casino rules. The dealer gives you chips of a unique color. You can use these chips only at the roulette table.

You bet by placing chips on the betting layout before the ball is dropped, or before the dealer announces "no more bets" soon after the

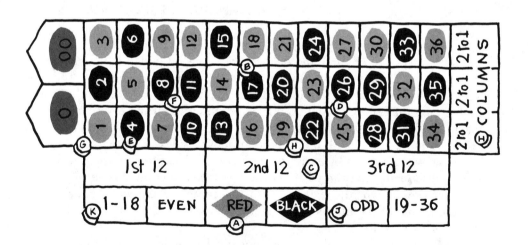

FIGURE 1

ball is dropped. You can bet on individual numbers, combinations of numbers, or colors. After each spin, the ball lands in a section, and bets are settled. In addition to describing roulette bets, I will tell you what happens if you play repeatedly. Assume for now that roulette wheels are evenly balanced, that is, the ball has the same chance of landing in every section (1 in 38). Here are three roulette bets.

> **Bet on red**—If you put chips in the box marked "RED" on the betting layout (A in Figure 1), and the ball lands in a red section, you win. If the ball lands in any other section (black or green), you lose. Eighteen of the thirty-eight sections are red, so the probability of winning a red bet equals 18 in 38: $P(\text{red}) = 18/38$.

> **Bet on #17**—If you put chips in the box marked "17" on the betting layout (B in Figure 1), and the ball lands in section 17 of the wheel, you win. If the ball lands in any other section, you lose. There is only one section numbered 17, so the probability of winning equals 1 in 38: $P(\#17) = 1/38$.

> **Bet on a number from 13 through 24**—This bet is called "2nd 12" or "middle 12." If you put chips in the box marked "2nd 12" on the betting layout (C in Figure 1), you win if the ball lands in any of the sections marked 13–24 and lose otherwise. Twelve of the sections are numbered from 13 to 24. The chance of winning equals 12 in 38: $P(\text{2nd }12) = 12/38$.

PAYOFF ODDS

Which of the three bets is best? You might say red, since the chance of winning a red bet is the highest. There's more to the story. In investments and gambling games, the higher the risk, the higher the profit if you win. If you invest in a savings account, you will almost surely get interest payments. If you gamble on a risky mining venture, the chance of success is small, but if you win, you win big.

In a casino, every bet has payoff odds. I'll express payoff odds in the form "X to Y," where X is the amount received in winnings for every Y dollars bet. For example, payoff odds of 4 to 1 means you get back $4 in winnings for every dollar bet. If you bet $1 and win, you get back a total of $5: $4 in winnings plus the $1 you bet. Payoff odds of 3 to 2 means that if you win, you get back $3 in winnings for every $2 you bet. It's convenient if $Y = 1$, so odds are expressed as winnings per dollar bet.

The payoff odds for a bet on red in roulette are 1 to 1, or *even* odds. If you bet a dollar and red comes up, you get back $1 plus the dollar you bet. The payoff odds for a bet on #17 are 35 to 1. If you bet $2 on #17 and win, you get back $70 plus the $2 you bet. The payoff odds for a bet on 2nd 12 are 2 to 1. If you bet $5 on 2nd 12 and a number from 13 through 24 comes up, you get back $10 plus the $5 you bet.

So which bet is best? One answer is, "It depends." Maybe you like the joy of winning, regardless of how much you win. In this case, you should bet on red, since of the three bets a red bet gives you the greatest chance of winning. Maybe you like to win, but only when you win big. Then you should bet on number 17. Or maybe you like to win *and* like to win big. You should bet on the middle 12 as a compromise. People make decisions based on personal preferences.

. . . .

I am beginning to remember also that the middle twelve numbers, to which I had become positively attached, turned up most frequently of all. There was a sort of pattern—they appeared three or four times running, without fail, then disappeared for two turns, then again appeared three or four times in succession. This remarkable regularity occurs sometimes in streaks—and this is what throws out the inveterate gamblers, always doing sums with a pencil in their hands. And what terrible jests fate sometimes plays!

—FYODOR DOSTOYEVSKI, *THE GAMBLER*

THE LAW OF AVERAGES AND THE HOUSE EDGE

Since there are 18 red sections on the wheel, the law of averages implies that in the long run, for every 38 spins, red will come up an average of 18 times and won't come up an average of 20 times. For any particular spin, it's uncertain whether or not red will come up.

Recall from Chapter 1 that the house edge is the casino's long-run average winnings per dollar bet, expressed as a percentage. If you bet a dollar on red repeatedly, you'll win an average of 18 bets out of 38 and lose an average of 20, for a $2 net loss. This gives an average loss of $2/38 = 5.3 cents per dollar bet. This makes the house edge 5.3%.

How about bets on #17? In repeated play, in 38 spins, #17 will come up an average of one time and won't come up the other 37 times. If you bet $1 each time, you'll win an average of once every 38 spins, with payoff $35 (payoff odds are 35 to 1), and lose $37 from 37 losing spins, for a net loss of $2. This gives an average loss of $2/38 = 5.3 cents per dollar bet. Again, the house edge equals 5.3%.

If you repeatedly bet a dollar on middle 12, every 38 spins you'll win an average of $24 (12 wins at 2 to 1 payoff odds) and lose an average of 26 spins, for a net loss of $2. This gives (you guessed it) an average loss of $2/38 = 5.3 cents per dollar bet. The house edge equals 5.3%.

In repeated play, the three bets are the same. They all give the house an edge of 5.3%. If you play long enough, you'll eventually start losing, your losses averaging 5.3 cents per dollar bet.

EXPECTED WINNINGS

There's a simple formula for computing a player's average winnings (or losses) for repeated play in a game like roulette. This mathematical average is called the player's "expected winnings."

Let's analyze a $1 bet on red. You either win $1, with probability 18/38, or you lose $1, with probability 20/38. To make it easier to visualize these probabilities, first construct a probability distribution table, where "1" in the win column denotes a $1 win and "−1" denotes a $1 loss. Here is the probability distribution table for a $1 bet on red:

$1 BET ON RED
. . . .

Win	Probability
1	18/38
−1	20/38

. . . .

To find the expected winnings, which I'll denote by E, multiply the win values by their probabilities and add, to get the weighted average

$$E = [1 \times (18/38)] + [(-1) \times (20/38)] = -2/38 = -.053$$

You can expect to win −.053, or lose an average of 5.3 cents per dollar bet on red. Converted to a percentage and dropping the minus sign, your expected winnings for a $1 bet becomes the house edge: 5.3%.

What happens to expected winnings if you bet $5 instead of $1? For a $5 bet on red, you either win $5, with probability 18/38, or lose $5, with probability 20/38. The probability distribution table and expected winnings are as follows:

$5 BET ON RED

Win	Probability
5	18/38
−5	20/38

$$E = [5 \times (18/38)] + [-5 \times (20/38)]$$
$$= 5 \times (-2/38) = -10/38$$

Expected winnings for a $5 bet equal five times expected winnings for a $1 bet. This is true in general: Expected winnings for a $Y bet equal Y times expected winnings for a $1 bet. Likewise, to find expected winnings for a $1 bet, divide expected winnings by the total amount bet. This is useful in computing house edge, since the house edge is always expressed as the house's expected winnings for a $1 bet.

Expected winnings can also be computed for gambling games that are not played in casinos. For example, insurance companies are interested in their expected profit (winnings) for a given policy. (See Problem 15 in "Figuring the Odds," Chapter 2, at the back of this book.)

Here are the probability distribution table, expected winnings, and house edge for $1 bets on #17 and 2nd 12:

$1 BET ON #17

Win	Probability
35	1/38
−1	37/38

$$E = [35 \times (1/38)] + [-1 \times (37/38)]$$
$$= -2/38$$
$$= -.053$$

House edge = 5.3%

$1 BET ON 2nd 12

Win	Probability
2	12/38
−1	26/38

$$E = [2 \times (12/38)] + [-1 \times (26/38)]$$
$$= -2/38$$
$$= -.053$$

House edge = 5.3%

Expected winnings is a useful number because it gives the actual average results of repeated play. As I remarked earlier, people often make choices based on personal preferences. Someone who is desperate for money will take more risks than someone with a steady job. An extra thousand in the bank means more to a poor person than to the Sultan of Brunei. Losses may be acceptable to someone who finds gambling entertaining. What's unfortunate is when someone gambles without knowing the financial consequences. I'll use expected winnings as a measure of financial consequences for many of the games we discuss. If you have your own reasons for gambling, don't let the prospect of losing money stop you.

CHURN

Often, gamblers on a winning streak will bet big and go broke quickly. They justify big bets when they're ahead because they're "playing with the house's money." Please. Money you win is *your* money. Rebetting winnings is called "churn" by casinos. Churn makes casino owners sleep easy. There aren't many gamblers who quit when they're ahead.

FAIR ODDS

Payoff odds for a bet are determined by the casino. For example, recall that the payoff odds of 35 to 1 for a bet on a single number result in a house edge of 5.3% and expected winnings of −5.3 cents per dollar bet. "Fair odds," or "true odds," are the odds that make the house edge, and the player's expected winnings, zero.

To compute fair odds, find the win and loss probabilities and take the ratio of losses to wins. For example, the chance of winning a bet on #17 equals 1/38, and the chance of losing equals 37/38. The fair odds are the ratio of losses to wins: 37 to 1. To show that these odds balance the probabilities, compute the expected winnings when the fair odds of 37 to 1, instead of the casino payoff odds of 35 to 1, are used for a $1 bet on #17.

$1 BET ON #17 WITH FAIR ODDS PAYOFF

Win (fair odds)	Probability
37	1/38
−1	37/38

$$E = [37 \times (1/38)] + [-1 \times (37/38)] = 0$$

House edge = 0

Here are fair odds compared with casino payoff odds for the roulette bets we've discussed:

Bet	P(win)	P(lose)	Fair Odds	Casino Payoff Odds
#17	1/38	37/38	37 to 1	35 to 1
Red	18/38	20/38	20 to 18	1 to 1
2nd 12	12/38	26/38	26 to 12	2 to 1

For example, the fair odds for a bet on red are 20 to 18, or 20/18 to 1. For a $1 bet on red, a casino giving fair odds would pay $20/18, or $1.11, instead of $1. No casinos give fair roulette odds. Why should they?

SUMMARY OF ROULETTE BETS

Here is a summary of all roulette bets. The house edge is 5.3% for every bet except the five numbers bet, which is worse for the player.

Single Number Bets. You make a single number bet by placing chips in one of the boxes numbered 1 through 36, 0, or 00 on the betting layout. A bet on #17 is a single number bet. Payoff odds are 35 to 1. Here are the probability distribution table, expected winnings, and house edge for single number bets.

$1 SINGLE NUMBER BET

Win	Probability
35	1/38
−1	37/38

$$E = [35 \times (1/38)] + [-1 \times (37/38)]$$
$$= -2/38$$
$$= -.053$$

House edge = 5.3%

Two Numbers Bets. You make a two numbers bet by placing chips on a boundary line between two number boxes (D in Figure 1). For example, you can make a bet on numbers 25 and 26, 11 and 8, or 1 and 0. If two numbers don't have a common boundary, you can't make this bet. For example, you can't make a two numbers bet on 9 and 17. The probability of winning a two numbers bet equals 2/38. Payoff odds are 17 to 1. Here are the probability distribution table, expected winnings, and house edge.

$1 TWO NUMBERS BET

Win	Probability
17	2/38
−1	36/38

$$E = [17 \times (2/38)] + [-1 \times (36/38)]$$
$$= -2/38$$
$$= -.053$$

House edge = 5.3%

Three Numbers Bets. You make a three numbers bet by placing chips on the outside boundary of any row of three numbers (E in Figure 1). For example, to bet on the numbers 4, 5, and 6, place chips on the outside boundary of the 4 box or 6 box. You can't make a three numbers bet on numbers that aren't in a line in one of the 12 rows on the betting layout, except for 0, 1, 2; 0, 2, 00; or 2, 00, 3.

    The probability of winning a three numbers bet equals 3/38. Payoff odds are 11 to 1.

$1 THREE NUMBERS BET

. . . .

Win	Probability
11	3/38
−1	35/38

. . . .

$$E = [11 \times (3/38)] + [-1 \times (35/38)]$$
$$= -2/38$$
$$= -.053$$

House edge = 5.3%

Four Numbers Bets. You make a four numbers bet by placing chips on the intersection of any square of four numbers (F in Figure 1). For example, you can bet on the four numbers 7, 8, 10, and 11, or 25, 26, 28, and 29. You can only make four numbers bets on four numbers with a common intersection. The probability of winning a four numbers bet equals 4/38. Payoff odds are 8 to 1.

$1 FOUR NUMBERS BET

. . . .

Win	Probability
8	4/38
−1	34/38

. . . .

$$E = [8 \times (4/38)] + [-1 \times (34/38)]$$
$$= -2/38$$
$$= -.053$$

House edge = 5.3%

Five Numbers Bet. There is only one five numbers bet, 0, 00, 1, 2, 3, which you make by putting chips on the outside intersection of the 1 and 0 sections (G in Figure 1). The probability of winning the five numbers bet equals 5/38. The payoff odds are 6 to 1. The house edge for the five numbers bet is an unhealthy 7.9%.

$1 FIVE NUMBERS BET
. . . .

Win	Probability
6	5/38
−1	33/38

. . . .

$$E = [6 \times (5/38)] + [-1 \times (33/38)]$$
$$= -3/38$$
$$= -.079$$

House edge = 7.9%

Six Numbers Bets. A six numbers bet is a bet on two adjacent rows of numbers (H in Figure 1). You make this bet by placing chips on the outside boundary of the intersection of the outside numbers of the two rows. For example, to bet on the numbers 19, 20, 21, 22, 23, and 24, place chips on the outside intersection of 19 and 22 or 21 and 24. The probability of winning a six numbers bet equals 6/38. Payoff odds are 5 to 1.

$1 SIX NUMBERS BET
. . . .

Win	Probability
5	6/38
−1	32/38

. . . .

$$E = [5 \times (6/38)] + [-1 \times (32/38)]$$
$$= -2/38$$
$$= -.053.$$

House edge = 5.3%

Dozens Bets. There are three dozens bets: 1st 12, 2nd 12, and 3rd 12 (C in Figure 1). You make a dozens bet by placing chips in the appropriate box on the betting layout. You win if the ball lands on any of the dozen numbers you bet on, and lose otherwise. The probability of winning a dozens bet equals 12/38. Payoff odds are 2 to 1.

$1 DOZENS BET

· · · · · · · · · · ·

Win	Probability
2	12/38
−1	26/38

· · · · · · · · · · · · ·

$$E = [2 \times (12/38)] + [-1 \times (26/38)]$$
$$= -2/38$$
$$= -.053$$

House edge = 5.3%

Column Bets. You make a column bet by placing chips in one of the three column boxes on the betting layout (I in Figure 1). You win if the ball lands on any of the numbers in the selected column. Since there are 12 numbers in each column, the probability of winning a column bet equals 12/38. Payoff odds are 2 to 1. The probability distribution table, expected winnings, and house edge (5.3%) are the same as for dozens bets.

Red and Black Bets. You bet on a red or black number by placing chips in the appropriate box on the betting layout (A in Figure 1). Since there are 18 red numbers and 18 black numbers, the probability of winning either of these bets equals 18/38. Payoff odds are 1 to 1.

$1 BET ON RED OR BLACK

· · · · · · · · · ·

Win	Probability
1	18/38
−1	20/38

· · · · · · · · · · · ·

$$E = [1 \times (18/38)] + [-1 \times (20/38)]$$
$$= -2/38$$
$$= -.053$$

House edge = 5.3%

Even and Odd Bets. You bet on an even or odd number (0 and 00 are excluded) by placing chips in the appropriate box on the outside of the betting layout (J in Figure 1). Since there are 18 even numbers and 18 odd numbers, the probability of winning either of these bets equals 18/38. Payoff odds are 1 to 1. The probability distribution table, expected winnings, and house edge (5.3%) for even and odd bets are the same as for red and black bets.

Bets on 1−18 and 19−36.

You bet that the ball will land on one of the numbers 1−18 or 19−36 by placing chips in the appropriate box on the betting layout (K in Figure 1). Since there are 18 numbers from 1−18 and 18 numbers from 19−36, the probability of winning either of these bets equals 18/38. Payoff odds are 1 to 1. The probability distribution table, expected winnings, and house edge (5.3%) for bets on 1−18 and 19−36 are the same as for red and black bets.

CAN YOU BEAT THE HOUSE EDGE?

Combination Bets.

Individual bets give the house a 5.3% edge. If you make one of these bets repeatedly, you are certain to eventually lose, at the rate of 5.3 cents per dollar bet. What about making a combination of bets? By now you might be able to guess that the house edge on combination bets is exactly the same as the house edge on individual bets: 5.3%. In fact, the expected winnings of a combination bet equals the sum of the expected winnings of the individual bets in the combination. If you're willing to accept this on faith, skip to the section on double-up strategies. If you want proof, read on.

Suppose you bet $1 on #17 (35 to 1 payoff odds), $2 on 1st 12 (2 to 1, and $5 on black (1 to 1), all on the same spin.

If the ball lands on #17, a black section, you win $35 for the bet on #17, win $5 for the bet on black, and lose $2 for the bet on 1st 12, for winnings of $40 − $2 = $38: P(you win $38) = 1/38.

If the ball lands on a black section numbered from 1 through 12, you win $5 for the bet on black, win $4 for the bet on 1st 12, and lose $1 for the bet on #17, for net winnings of $8. There are 6 black sections between 1 and 12: P(you win $8) = 6/38.

If the ball lands on a black section numbered from 13 through 36 but not 17, you win $5 for the black bet, lose $1 for the bet on #17, and lose $2 for the bet on 1st 12, yielding net winnings of $2. There are 11 such sections, so P(you win $2) = 11/38. This is tedious.

If the ball lands on a red section numbered from 1 through 12, you win $4 for the bet on 1st 12, you lose $1 for the bet on #17, and you lose $5 for the bet on black, for a net loss of $2. Since there are 6 red sections in 1st 12, P(you lose $2) = 6/38.

If the ball lands on 0, 00, or a red section numbered from 13 through 36, you lose $8, everything you bet. Since there are 14 such sections, P(you lose $8) = 14/38.

Here are the probability distribution table and your expected winnings for this combination bet. Note that the probabilities in the distribution table sum to 1.

COMBINATION BET

Win	Probability
38	1/38
8	6/38
2	11/38
−2	6/38
−8	14/38

$$E = -\$16/38$$

To find the house edge, divide expected winnings (−$16/38) by the amount bet ($8), drop the minus sign, and convert to a percentage. Guess what?

$$\text{House edge} = 5.3\%$$

We now compute expected winnings = −$16/38 for the combination bet without going through the hassle of finding the probability distribution table. All we do is add the expected winnings of the individual bets:

$1 BET ON #17		$2 BET ON 1ST 12		$5 BET ON BLACK	
Win	Probability	Win	Probability	Win	Probability
35	1/38	4	12/38	5	18/38
−1	37/38	−2	26/38	−5	20/38
$E = -2/38$		$E = -4/38$		$E = -10/38$	

$$E(\text{combination bet}) = (-2/38) + (-4/38) + (-10/38) = -16/38$$

Bottom line: You can't combine a group of bad bets into a single good bet.

Double-up Strategy.

I once had a discussion with Nick, the doorman for a Manhattan apartment building. "Progression is the key," said Nick. "If you lose, double your bet to cover your losses. Eventually, you'll win. Quit when you're ahead." This strategy is known as the "double-up," or "martingale," system.

Here is a version of the double-up strategy.

1st bet—Bet $1 on red. If red comes up, you win $1. Quit.

2nd bet (if necessary)—If red doesn't come up, double your bet and bet $2 on red. If red comes up, you win $2, covering your previous loss and leaving you a $1 profit. Quit.

3rd bet—If you lose the second bet, double your bet and bet $4 on red. If red comes up, you win $4, covering your previous $1 and $2 losses and leaving you a $1 profit. Quit.

4th bet—If you lose the third bet, double your bet and bet $8 on red. If red comes up, you win $8, covering your $1, $2, and $4 losses, and leaving you a $1 profit. Quit.

5th bet—If you lose the fourth bet, double your bet and bet $16 on red. If red comes up, you win $16, covering your $1, $2, $4, and $8 losses, leaving you a $1 profit. Quit.

. . . etc.

Red is certain to eventually come up. At that time, you quit a winner, with a $1 profit. Is this a guaranteed way to beat the law of averages? Two problems arise.

In roulette as in other casino games, there is a house limit, both a minimum and maximum bet. For example, at Caesar's Palace in Lake Tahoe, Nevada, one roulette table has a $1 minimum and a $1000 maximum. You can bet more on "outside" bets with low payoff odds (like red and black) than "inside" bets with high payoff odds (like single number bets).

Why is there an upper limit if the casino has an edge? Some people have more money than a casino. If the Sultan of Brunei comes in and bets $50 million dollars on red, one spin of the wheel could put the casino out of business. The Sultan of Brunei has too much leverage. The casino doesn't want to risk going broke from one unlucky spin when the law of averages guarantees success in the long run with moderate betting.

You probably don't have as much money as the Sultan of Brunei. If you run into an unlucky losing streak, you may not be able to bet enough to cover your losses. For example, if red fails to come up 15 times in a row, on the sixteenth bet you must bet $32,768 in an attempt to come out $1 ahead.

What does it mean to quit? Most likely, red will come up within a few spins, and you will win $1. Does quitting mean you should go home and never return to the casino? Or does it mean you should have a drink and then start the betting cycle again?

If you use the double-up cycle repeatedly, each time starting with a $1 bet, you will win most of the time, making a $1 profit each time. Unfortunately, you will eventually encounter a sequence of unlucky spins and lose big, erasing all your $1 gains.

When I explained this to Nick the doorman, he told me that the way to avoid devastating unlucky streaks was to change casinos before they occurred. When I asked Nick how he could tell in advance when he was going to have an unlucky streak, he told me that experience was the key.

Let's analyze a double-up betting cycle, computing expected winnings and house edge. To do this, we use what is known as the multiplication rule for independent events. Spins of a roulette wheel, like coin tosses and dice rolls, are independent events. That means the outcome of one spin has no effect on the outcome of any other spin.

> The multiplication rule for independent events states that the probability of the occurrence of a sequence of independent events equals the product of the individual probabilities.

For example, suppose you make a sequence of red bets in roulette, where $P(\text{red}) = 18/38$ and $P(\text{not red}) = 20/38$. Letting W denote "win" and L denote "lose," we use the multiplication rule to calculate

$$P(\text{win first two bets}) = P(WW) = (18/38) \times (18/38) = .2244$$
$$P(\text{lose first two bets}) = P(LL) = (20/38) \times (20/38) = .2770$$
$$P(\text{lose 1st, win 2nd}) = P(LW) = (20/38) \times (18/38) = .2493$$
$$P(LLW) = (20/38)^2 \times (18/38) = .1312$$
$$P(LLLW) = (20/38)^3 \times (18/38) = .0691$$
$$P(LLLLW) = (20/38)^4 \times (18/38) = .0363$$
$$P(LLLLLW) = (20/38)^5 \times (18/38) = .0191$$
$$\ldots \text{ etc.}$$

The multiplication rule for independent events can be used to compute seemingly complicated probabilities, such as the probabilities in a double-up betting cycle.

Now, think of the entire double-up cycle as one complicated bet. Start with a $1 red bet. If you lose, double your bet. Keep doubling your bet until you win (red comes up). Then quit. The maximum bet is $1000. Suppose you have enough money to keep betting until you reach the maximum. The following table gives the relevant information. The probabilities are calculated using the multiplication rule for independent events. The betting cycle ends when red comes up for the first time or you bet the maximum.

DOUBLE-UP CYCLE

Starting with $1 Bet on Red ($1000 maximum bet)

Bet No.	Amount Bet	Total Losses Before Bet	Profit If Cycle Ends Here	Probability Cycle Ends Here
1	$1	0	$1	$18/38 = .4737$
2	2	1	1	$(20/38) \times (18/38) = .2493$
3	4	3	1	$(20/38)^2 \times (18/38) = .1312$
4	8	7	1	$(20/38)^3 \times (18/38) = .0691$
5	16	15	1	$(20/38)^4 \times (18/38) = .0363$
6	32	31	1	$(20/38)^5 \times (18/38) = .0191$
7	64	63	1	$(20/38)^6 \times (18/38) = .0101$
8	128	127	1	$(20/38)^7 \times (18/38) = .0053$
9	256	255	1	$(20/38)^8 \times (18/38) = .002789$
10	512	511	1	$(20/38)^9 \times (18/38) = .001468$
11	1000	1023	−23	$(20/38)^{10} \times (18/38) = .000773$
		or (if lose)	−2023	$(20/38)^{11} = .000858$

Although the chance of red not coming up 11 times in a row is less than .001, if this happens, you lose $2023. The chance is greater than .998 that red will come up within the first 10 bets. In this case, you quit with a $1 profit.

To compute your expected winnings with this betting system, multiply the profits (and losses) by the corresponding probabilities and add, to get

$$E = (1 \times .4737) + (1 \times .2493) + \ldots + (-2023 \times .000858)$$
$$= \$ -.76$$

Even though you win most of the time, occasional large losses cause an average loss of 76 cents per cycle. To find the house edge, divide the expected winnings by the total amount bet. Since the total amount bet depends on when you quit, use the expected, or average, total amount bet.

You find the average amount bet in a betting cycle as follows: Multiply the amount bet for each outcome of the cycle by the probability that the cycle ends on that bet and add. For example, the total amount bet equals $1 if red comes up on the first bet. If red comes up for the first time on the second bet, the total amount bet equals $1 + $2 = $3. If red comes up for the first time on the third bet, the total amount bet equals $1 + $2 + $4 = $7, etc. Looking at the double-up-cycle table, the total amount bet equals "total losses before bet" for the next bet in the cycle. Multiplying the outcomes by their probabilities and adding, we get

E(total amount bet)
$$= [1 \times (18/38)] + [3 \times (20/38) \times (18/38)] + \ldots = \$14.37$$

To get the house edge, divide expected winnings by expected total amount bet:

$$\frac{E\text{(winnings)}}{E\text{(total bet)}} = \frac{-\$.76}{\$14.37} = -.053$$

The house edge = 5.3%. You can't win!

WATCH THE WHEEL

Suppose you are following a popular roulette strategy I call "watch the wheel." Red comes up 10 times in a row. What should you do?

Here are three possible strategies. The one you pick answers the age-old riddle, "How do you tell the difference between a mathematician, a statistician, and a fool?"

1 – Black is "due." Bet on black.

2 – Red is "hot." Bet on red.

3 – It's a random fluke. Don't bet.

Here's the answer to the age-old riddle.

The mathematician knows that spins of a roulette wheel are independent events. The law of averages doesn't apply to any particular spin, and an evenly balanced wheel has the same chance of coming up red or black, even after 10 reds in a row. The mathematician chooses strategy 3: Don't bet.

The statistician observes that 10 reds in a row is unlikely to occur with an evenly balanced wheel. Therefore, the wheel may be unbalanced and biased toward red. The statistician chooses strategy 2: Bet on red.

The fool thinks black is due.

YOU CAN'T WIN

A winning roulette system is impossible. In repeated play, you're certain to lose, your long-run losses averaging 5.3 cents per dollar bet. If you play roulette, do it for a reason other than wanting to make money in repeated play.

CHAPTER

3

CRAPS

Ever since the big craps game, I been livin' on chicken and wine. I'm a
leader of society since I got mine.

—RY COODER, "CHICKEN SKIN MUSIC"

Craps is a fast-moving dice game played on a classy table with a complex
betting layout (Figure 2). The game is run by four casino employees: The
boxman (puts your money in a box), the stickman (rakes in the dice with a
stick), and two dealers (such a deal). Bettors stand around the table and
place chips on the betting layout. You buy chips from the dealer or
boxman.

Bets are made on outcomes of the dice rolls. Many bets are available.
A crowded craps game with screaming bettors is what gambling is all
about. Playing craps, you get a vivid picture of the law of averages, as you
watch your bankroll oscillate, slowly drifting toward zero.

One of the bettors, called the "shooter," rolls the dice. This neat job
rotates among bettors at the craps table. Don't be shy when it's your turn.
Shake the dice and fling them on the table, trying to hit the table wall
opposite you. It's especially fun to aim your dice at piles of chips bettors
have placed on the betting layout. Don't throw the dice so hard that they
sail off the table, though, or you'll irritate casino employees.

I was playing craps at the Mirage Casino in Las Vegas. The boxman,
who looked like a Chicago gangster during prohibition (many boxmen
look like this), was staring with glazed eyes at the betting layout. One of
the bettors threw the dice so hard they bounced off the table onto the
floor. The boxman, who didn't seem to notice, cleared his throat without
looking up. The dice were retrieved, and the shooter tossed them off the
table again. The boxman, still not looking up, said with an emotionless
voice, "You're not going to win any money on the floor, my friend."

Unlike roulette bets, almost all of which give the house a 5.3% edge,
craps bets offer a variety of house edges, some worse than roulette, some
better.

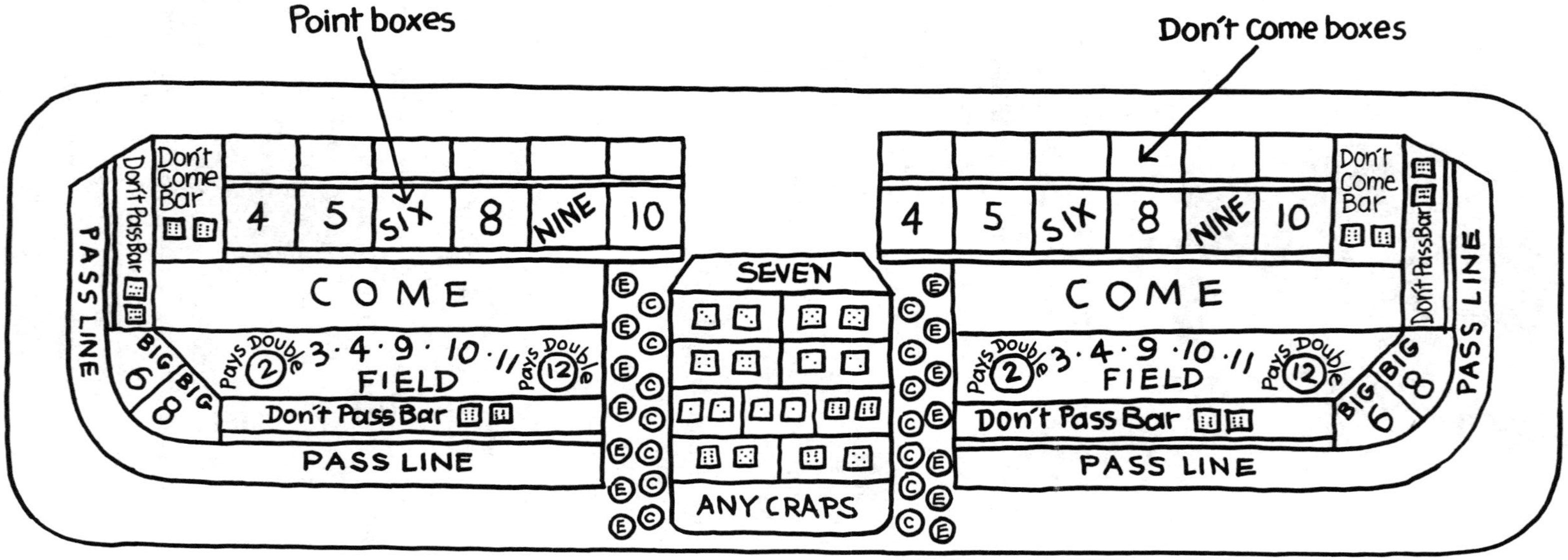
Point boxes
Don't come boxes
Don't Come Bar
Don't Pass Bar
4
5
SIX
8
NINE
10
COME
Pays Double 2
3 · 4 · 9 · 10 · 11
Pays Double 12
FIELD
BIG 6
BIG 8
Don't Pass Bar
PASS LINE
SEVEN
ANY CRAPS

FIGURE 2

Before analyzing craps bets, let's talk dice combinations.

DICE COMBINATIONS
AND PROBABILITIES

A die is a cube having six faces with one to six dots, and thus there are six possible outcomes for each toss of a die. Craps is played with a pair of dice, and after each throw, the sum of the upturned dots is counted.

How many sums, or combinations, are there when you roll a pair of dice? They range from 2, when both dice come up 1 ("snake eyes"), to 12, when both dice come up 6 ("boxcars"). Suppose a pair of dice, one of which is yellow (Y) and one of which is blue (B), is thrown. The following table shows the 36 possible combinations:

THE 36 DICE COMBINATIONS

(Y	B)	(Y	B)	(Y	B)	(Y	B)	(Y	B)	(Y	B)
1	1	2	1	3	1	4	1	5	1	6	1
1	2	2	2	3	2	4	2	5	2	6	2
1	3	2	3	3	3	4	3	5	3	6	3
1	4	2	4	3	4	4	4	5	4	6	4
1	5	2	5	3	5	4	5	5	5	6	5
1	6	2	6	3	6	4	6	5	6	6	6

Since there are six possible outcomes for each die, there are $6 \times 6 = 36$ combinations. If you roll a pair of evenly balanced dice, the chance of getting each of the 36 outcomes equals 1/36.

Since there is only one way to get 2 and one way to get 12, the probability of throwing either of these combinations is 1/36. What about the other possible outcomes? Let's take 7. Looking at the table, observe that there are six ways to get 7: 1 6, 6 1, 2 5, 5 2, 3 4, and 4 3. Each of these outcomes has probability 1/36. Using the addition rule of probability, which states that the probability of an event is the sum of the probabilities of the individual outcomes that make up the event, we find the chance that 7 will come up is

$$(1/36) + (1/36) + (1/36) + (1/36) + (1/36) + (1/36) = 6/36 = 1/6$$

The number of combinations for each dice sum and the corresponding probabilities are as follows.

Sum of Dots	No. of Combinations	Probability
2	1	1/36
3	2	2/36
4	3	3/36
5	4	4/36
6	5	5/36
7	6	6/36
8	5	5/36
9	4	4/36
10	3	3/36
11	2	2/36
12	1	1/36

The most likely result is 7, followed by 6 and 8, 5 and 9, 4 and 10, 3 and 11, 2 and 12. Now let's look at craps bets.

PASS LINE BET

The pass line bet is central in craps. It works like this:

The first roll of the dice is called the come-out roll. You make a pass line bet before the come-out roll by placing chips in the pass line area of the betting layout (Figure 2). Three things can happen on the come-out roll:

7 or 11 comes up. Pass line bettors win.

2, 3, or 12 comes up. This is known as "craps." Pass line bettors lose.

4, 5, 6, 8, 9, or 10 comes up. If one of these numbers, called "points," is rolled on the come-out roll, the shooter continues to roll the dice until one of two things happens: The point is matched before 7 comes up; pass line bettors win. 7 comes up before the point is matched; pass line bettors lose.

The rolls after the come-out roll until the point or 7 come up are called point rolls. During point rolls, the only outcomes that affect the pass line bet are the point and 7. The outcomes 2, 3, and 12, which cause pass line bettors to lose on the come-out roll, and 11, which causes pass line bettors to win on the come-out roll, don't count on point rolls.

For example, suppose the come-out roll is 8. The point is 8. For pass line bettors to win, 8 must be rolled before 7. Suppose the point rolls that follow are 9, 3, 11, 4, 5, and 8. Since 8 came up before 7, pass line bettors win. Suppose the come-out roll is 4, and the point rolls that follow are 5, 6, 12, 9, 9, and 7. Since 7 came up before 4, pass line bettors lose. As soon as the pass line bet is determined, the next roll of the dice is another come-out roll.

Two disks, called point markers, are used to indicate come-out and point rolls. If the current roll is a come-out roll, the disks are placed away from the numbered point boxes along the top of the betting layout. When a point is established, the disks are placed in the appropriate point boxes (Figure 3). You can tell from the point markers whether the next roll is a come-out roll or a point roll, and what the point is.

Pass line payoff odds are 1 to 1. The house edge for pass line bets equals 1.4%, considerably lower than the 5.3% roulette house edge.

$1 PASS LINE BET

. . . .

Win	Probability
1	.493
−1	.507

. . . .

$$E = (1 \times .493) - (1 \times .507) = -.014$$
$$\text{House edge} = 1.4\%$$

Point marker on 8

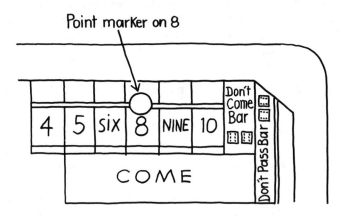

FIGURE 3

Here is why the chance of winning a pass line bet is .493, almost 50%. The calculation is complex. If you're not interested, skip ahead to the section on come bets.

You win a pass line bet either if you win on the come-out roll or win by matching a point. It follows from the addition rule for probability that

$$P(\text{win pass line bet}) = P(\text{win on come-out roll})$$
$$+ P(\text{win by matching point})$$

You win on the come-out roll if 7 or 11 comes up, so

$$P(\text{win on come-out roll}) = 8/36$$

To find the chance of winning by matching a point, consider each point separately:

$$P(\text{win by matching point}) = P(\text{win by matching 4})$$
$$+ P(\text{win by matching 5})$$
$$+ P(\text{win by matching 6})$$
$$+ P(\text{win by matching 8})$$
$$+ P(\text{win by matching 9})$$
$$+ P(\text{win by matching 10})$$

To win by matching 4, two things must occur: 4 is the result of the come-out roll, and then 4 comes up before 7. From the dice table, check that $P(4 \text{ is result of come-out roll}) = 3/36 = 1/12$. To find $P(4 \text{ comes up}$ before 7), note that there are six ways to get 7 and three ways to get 4, for a total of nine ways that determine the bet. If something else comes up, the shooter rolls again. Since three of the nine outcomes that determine the bet result in 4, $P(4 \text{ before } 7) = 3/9 = 1/3$. Since getting a 4 before 7 is independent of the result of the come-out roll, it follows from the multiplication rule for independent events that

$$P(\text{win by matching 4}) = P(4 \text{ on come-out roll}) \times P(4 \text{ before } 7)$$
$$= (1/12) \times (1/3)$$
$$= 1/36$$

To win by matching 5, 5 must be the result of the come-out roll, and then 5 must come up before 7. Check that $P(5 \text{ is result of come-out}$ roll$) = 4/36 = 1/9$. There are four ways to get 5, and six ways to get 7, for a total of $4 + 6 = 10$ ways to determine the bet. Thus, $P(5 \text{ before } 7)$ $= 4/10 = 2/5$, and

$$P(\text{win by matching 5}) = P(\text{5 on come-out roll}) \times P(\text{5 before 7})$$
$$= (1/9) \times (2/5)$$
$$= 2/45$$

To win by matching 6, 6 must be the result of the come-out roll, and then 6 must come up before 7. Check that $P(\text{6 is result of come-out roll}) = 5/36$. There are five ways to get 6, and six ways to get 7, for a total of eleven ways to determine the bet. Thus, $P(\text{6 before 7}) = 5/11$ and

$$P(\text{win by matching 6}) = P(\text{6 on come-out roll}) \times P(\text{6 before 7})$$
$$= (5/36) \times (5/11)$$
$$= 25/396$$

The points 8, 9, and 10 have the same number of combinations as 6, 5, and 4, respectively. Thus, by doubling the sum of the probabilities of winning by matching 4, 5, and 6, we get the overall chance of winning by matching a point:

$$P(\text{win by matching point}) = 2 \times [P(\text{win by matching 4})$$
$$+ P(\text{win by matching 5})$$
$$+ P(\text{win by matching 6})]$$
$$= 2 \times [(1/36) + (2/45) + (25/396)]$$

It follows that

$$P(\text{win pass line bet}) = P(\text{win on come-out roll})$$
$$+ P(\text{win by matching point})$$
$$= 8/36 + 2 \times [(1/36) + (2/45)$$
$$+ (25/396)]$$
$$= .493$$

Also,

$$P(\text{lose pass line bet}) = 1 - P(\text{win})$$
$$= 1 - .493$$
$$= .507$$

COME BET

The trouble with the pass line bet is that you only make it on a come-out roll. During sequences of point rolls, pass line bettors have nothing to do. This is not good. The bettors like action, and the casino doesn't like bettors who aren't betting. The come bet allows the bettor to make the equivalent of a pass line bet on every roll of the dice.

If the current roll is a point roll, and you want to make a pass line bet, place chips in the COME area of the betting layout (Figure 2). If you make a come bet, then, for you, the next roll of the dice is the same as a come-out roll on a pass line bet. Here's an example.

Suppose the come-out roll is 9. The point is 9, and the point markers are moved to 9 on the betting layout. You arrive at the table and make a come bet. The next roll is 6. This makes 6 your point. If 6 comes up before 7, you win. If 7 comes up before 6, you lose. The pass line bettors' point is still 9.

To keep track of your come bet, the dealer puts your chips in the 6 point box. Suppose the next roll is 9. Pass line bettors win, but since your point is 6, your come bet is undetermined. Since pass line bets were determined, the next roll is a come-out roll. For your bet, however, the next roll is a point roll. Suppose the next roll is 7. Pass line bettors win, since 7 came up on a come-out roll, but you lose, since 7 came up before your point of 6. On the other hand, suppose the next roll is 8. This makes the point 8 for new pass line bettors. The point for your come bet remains 6. If 6 comes up before 7, you win. If 7 comes up before 6, you lose.

Confused? Don't worry. Your chances of winning are the same, whether you understand what's happening or not.

Since the bets are identical, the payoff odds (1 to 1), win probability (.493), and house edge (1.4%) are the same for the come bet as for the pass line bet.

DON'T PASS BET

The don't pass bet is the opposite (almost) of the pass line bet and allows the bettor to bet against the pass line. You make a don't pass bet by placing chips in the DON'T PASS area of the betting layout (Figure 2). A don't pass bet can only be made on a come-out roll and is determined as follows.

Come-out roll: 7 or 11 — lose
2 or 3 — win
12 — tie

Point roll: Point before 7 — lose
7 before point — win

Thus, don't pass bets win whenever pass line bets lose, except when 12 comes up on the come-out roll, which is a tie. The reason for a tie is so the house can still have an edge. (In some casinos, 2 rather than 12 on the come-out roll is a tie.) Payoff odds for don't pass bets are 1 to 1.

You lose a don't pass bet only if pass line bets win. Since the chance of winning a pass line bet is .493, it follows that

$$P(\text{lose don't pass bet}) = .493$$

You tie a don't pass bet only if 12 comes up on the come-out roll. Thus, $P(\text{tie}) = 1/36 = .028$. You win a don't pass bet only if pass line bets lose and 12 isn't the result of the come-out roll. Since the chance is .507 that pass line bets lose, it follows that

$$P(\text{win don't pass bet}) = .507 - .028 = .479$$

Here is the house edge for don't pass bets.

$1 DON'T PASS BET
. . . .

Win	Probability
1	.479
0	.028
−1	.493

. . . .

$$E = (1 \times .479) + (0 \times .028) - (1 \times .493) = -.014$$
$$\text{House edge} = 1.4\%$$

The house edge for don't pass bets is the same as for pass line bets!

DON'T COME BET

The don't come bet is the same as the don't pass bet, except it is made on a point roll. In other words, the don't come bet is to the don't pass bet as the come bet is to the pass line bet. That's easy for me to say. The don't come bet allows you to make a bet equivalent to a don't pass bet whenever you want. You make a don't come bet by placing chips in the DON'T COME area of the betting layout (Figure 4).

For example, suppose the point is 9 and you make a don't come bet. Say the next roll is 5. Your point is 5, and your chips are moved to the upper sector of point box 5 (Figure 4). Since you made a don't come bet, you win if 7 comes up before 5. If the next rolls are 6, 4, 3, 9, and 7, pass line bettors win (the pass line point of 9 came up before 7), you win (7

Initial Don't Come bet

FIGURE 4

came up before 5), and those who made come bets when you made your don't come bet lose. Since the 9 before the 7 was the point, the 7 that determined your don't come bet was a come-out roll. Everyone who made a pass line bet on that roll won.

Are you getting a headache? Relax. I'll tell you now about odds bets, which when played in combination with the bets we've just discussed, reduce the house edge to less than 1%.

ODDS BETS

If a point is rolled after you make a pass, don't pass, come, or don't come bet, you can make an additional bet called an odds bet. The odds bet has to follow your initial bet. If you win, you are paid fair odds for the odds bet. If you could make an odds bet alone and not as a backup bet, the house edge would be zero. Unfortunately, since you can only make odds bets in combination with other bets, the house still maintains an edge. An odds bet combined with a pass, don't pass, come, or don't come bet makes the house edge lower than 1%.

Odds Bets and Pass Line Bets. After making a pass line bet, you can make an odds bet when a point is established. To do this, place chips on the betting layout directly behind your pass line bet, as in Figure 5. The amount you can bet varies. In many casinos you can bet up to twice the pass line bet. This is known as "double odds." Some casinos offer triple odds. You win an odds bet whenever you win the pass line bet. Payoff

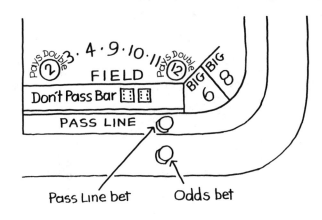

FIGURE 5

odds for the odds bet are fair odds, determined by the probability of winning. For example, if you make an odds bet to back up a pass line bet when the point is 4, since $P(4 \text{ before } 7) = 1/3$, payoff odds are fair odds of 2 to 1.

Suppose you make a $10 pass line bet, and the come-out roll is 4. You then make a $20 odds bet by placing $20 behind your pass line bet. If 7 comes up before 4, you lose both bets. If 4 comes up before 7, you win $10 from the pass line bet (payoff odds are 1 to 1) and $40 from the odds bet (payoff odds when the point is 4 are 2 to 1). You win a total of $50.

Odds Bets and Don't Pass Bets.

You can back up a don't pass bet with an odds bet once a point is established. You specify your bet to the dealer, who places your chips next to the don't pass bet. You win the odds bet whenever you win the don't pass bet (7 comes up before the point). In this case, because 7 is more likely to come up first than the point, the fair payoff odds are lower than the 1 to 1 don't pass payoff odds. For example, if you make an odds bet after a don't pass bet and 4 is the point, $P(\text{win odds bet}) = P(7 \text{ before } 4) = 2/3$, and payoff odds are 1 to 2.

Suppose you make a $6 don't pass bet and the come-out roll is 8. You then make a $12 odds bet to back up your don't pass bet. If 8 comes up before 7, you lose both bets. If 7 comes up before 8, you win $6 from the don't pass bet (payoff odds are 1 to 1) and $10 from the odds bet (payoff odds when the point is 8 are 5 to 6). You win a total of $16.

Odds Bets with Come and Don't Come Bets.

Odds bets can also be made after come bets and don't come bets when a point is established. You give chips to the dealer, who puts them by your original bet. As with pass and don't pass odds bets, payoff odds are fair odds.

Here are the payoffs and win probabilities for odds bets.

ODDS BETS WITH PASS LINE AND DON'T PASS BETS

. . . .

	WIN PROBABILITIES		PAYOFF ODDS	
Point	Pass Line P(point before 7)	Don't Pass P(7 before point)	Pass	Don't Pass
4	1/3	2/3	2 to 1	1 to 2
5	2/5	3/5	3 to 2	2 to 3
6	5/11	6/11	6 to 5	5 to 6
8	5/11	6/11	6 to 5	5 to 6
9	2/5	3/5	3 to 2	2 to 3
10	1/3	2/3	2 to 1	1 to 2

. . . .

House Edge. To find the house edge for odds bet combinations, divide expected winnings by total amount bet to obtain expected winnings per dollar bet, drop the minus sign, and convert to a percentage. This is trickier than for other bets we've discussed, because you can only make an odds bet when a point is established, and this happens randomly. Here is an informal explanation.

When you make a pass line, don't pass, come, or don't come bet, the house edge equals 1.4%. Expected winnings for a $1 bet equals −$.014. Since you get fair odds for an odds bet, expected winnings for an odds bet equals zero. Thus, total expected winnings for an odds bet combined with a $1 pass line, don't pass, come, or don't come bet equals −$.014 + 0 = −$.014.

Suppose you make $1 pass line bets and $1 odds bets whenever possible. A point is established if 4, 5, 6, 8, 9, or 10 is the result of a come-out roll. Refer to the dice table and verify that there are three ways to get a 4 or a 10, four ways to get a 5 or a 9, and five ways to get a 6 or an 8, for a total of twenty-four combinations that establish a point. Thus, P(establish point) = 24/36 = 2/3.

Since P(establish point) = 2/3, your combined bet will be $2 ($1 pass line and $1 odds) about two-thirds of the time and $1 (can't make the odds bet) about one-third of the time, for an average total amount bet of [$2 × (2/3)] + [$1 × (1/3)] = $5/3. Since your expected winnings equals −$.014, your expected winnings per dollar bet equals −$.014/(5/3) = −$.0084, giving a house edge of .84%.

If you make $1 pass line bets and double odds bets whenever possible, your combined bet will be $3 ($1 pass line and $2 odds) about two-thirds of the time and $1 (can't make the odds bet) about one-third of the time, for an average total bet of [$3 × (2/3)] + [$1 × (1/3)] = $7/3.

Because of fair odds, your combined expected winnings are still −$.014, and your expected winnings per dollar bet equals −$.014/(7/3) = −$.0060. The house edge is reduced to .60%.

ONE ROLL BETS

"One roll" bets are determined by the next roll of the dice after you make your bet. They are easy to analyze: They give the house an enormous edge.

Bet on 7. By placing chips in the box marked **SEVEN** in the center of the betting layout (Figure 2), you bet that the next roll of the dice will be 7. If 7 comes up on the next roll, you win. Otherwise, you lose. The payoff odds are 4 to 1. Since P(7 comes up) = 1/6, and P(7 doesn't come up) = 5/6, the house edge is as follows.

$1 BET ON 7
. . . .

Win	Probability
4	1/6
−1	5/6

. . . .

$$E = [4 \times (1/6)] - [1 \times (5/6)] = -1/6 = -.167$$
$$\text{House edge} = 16.7\%$$

Oy vey! The payoff odds of 4 to 1 (fair odds are 5 to 1) give the casino a huge edge.

Bets on 2, 3, 11, or 12. By placing chips in the appropriate box in the center of the betting layout (Figure 2), you can bet on any of the individual numbers 2, 3, 11, or 12. The house edge for these bets is 16.7%, the same as betting on 7. Bets on 2 or 12 (separate bets) have win probabilities P(win) = 1/36, and payoff odds 29 to 1. Bets on 3 or 11 (separate bets) have win probabilities P(win = 2/36), and payoff odds of 14 to 1. The house edges are as follows.

$1 BET ON 2 (OR 12)
. . . .

Win	Probability
29	1/36
−1	35/36

. . . .

$$E = [29 \times (1/36)] - [1 \times (35/36)]$$
$$= -6/36$$
$$= -.167$$

House edge = 16.7%

$1 BET ON 3 (OR 11)
. . . .

Win	Probability
14	2/36
−1	34/36

. . . .

$$E = [14 \times (2/36)] - [1 \times (34/36)]$$
$$= -6/36$$
$$= -.167$$

House edge = 16.7%

The house edge for all one number, one roll bets is 16.7%. These are terrible bets. You might as well go to the carnival and try to win a teddy bear by flipping a dime on a plate.

Any Craps Bet. "Craps" is slang for 2, 3, or 12 on a come-out roll. You can bet that 2, 3, or 12 will be the result of the next roll (not necessarily a come-out roll) by placing chips in the **ANY CRAPS** box in the center of the betting layout (Figure 2). If 2, 3, or 12 comes up on the next roll, you win. If anything else comes up, you lose. Since $P(2 \text{ comes up}) = 1/36$, $P(3 \text{ comes up}) = 2/36$, and $P(12 \text{ comes up}) = 1/36$,

$$P(\text{win any craps bet}) = (1/36) + (2/36) + (1/36) = 4/36 = 1/9$$

Payoff odds are 7 to 1. (Fair odds are 8 to 1.) The house edge is 11.1%. Don't make this bet unless advised to do so by a reliable astrologer.

$1 ANY CRAPS BET
. . . .

Win	Probability
7	1/9
−1	8/9

. . . .

$$E = [7 \times (1/9)] - [1 \times (8/9)] = -1/9 = -.111$$

House edge = 11.1%

Field Bet.

You make a field bet by placing chips in the **FIELD** box on the betting layout (Figure 2). If the result of the next roll is 3, 4, 9, 10, or 11, you win and are paid even money (1 to 1 payoff odds). If 2 or 12 comes up, you are paid double (2 to 1). If the result is anything else (5, 6, 7, or 8), you lose. This bet has a house edge of 5.6%. Using the dice table, we find P(3, 4, 9, 10, or 11) = 14/36, P(2 or 12) = 2/36, and P(5, 6, 7, or 8) = 20/36. The house edge is computed as follows.

$1 FIELD BET
· · · · · · · · · ·

Win	Probability
2	2/36
1	14/36
−1	20/36

· · · · · · · · · · · ·

$$E = [2 \times (2/36)] + [1 \times (14/36)] - [1 \times (20/36)]$$
$$= -2/36 = -.056$$

House edge = 5.6%

In some casinos, the field bet payoff odds are 3 to 1 if either 2 or 12 (not both) comes up. In this case, the house edge is reduced to 2.8%.

Craps-Eleven (Horn) Bet.

A craps-eleven bet is made by placing chips on one of the circled C's or E's in the center of the betting layout (Figure 2). You win a craps-eleven bet if the next roll is 2, 3, 11, or 12 (craps or 11) and lose otherwise. The dice table shows that there are six ways to win, so P(win craps-eleven bet) = 6/36 = 1/6. The payoff odds vary from casino to casino. The house edge is always over 12%. Enough.

· ·

You can't really understand the value of money until you go bankrupt at least twice.

—VETERAN LAS VEGAS GAMBLER

PLACE BETS

You win a pass line bet on a point roll if the point comes up before 7. With a place bet you can bet that a point comes up before 7, independently of the pass line bet. You make a place bet by giving chips to the dealer and specifying which point you want to bet on. The dealer will "place" your chips on the border of the appropriate point box.

There is no come-out roll with a place bet. For example, if you want to make a place bet on 8, you give chips to the dealer and say, "place bet on 8." Starting with the next roll, if 8 comes up before 7, you win. If 7 comes up before 8, you lose.

When analyzing pass line bets, we found the chance of a point coming up before 7. Place bet probabilities are similar. You win a place bet on 4 if 4 comes up before 7. If 7 comes up before 4, you lose. When anything else comes up, your place bet isn't affected. There are three ways to get 4 and six ways to get 7, for a total of nine outcomes that determine a place bet on 4. Since three of these nine outcomes win, P(win place bet on 4) = 3/9 = 1/3. You compute win probabilities for the other place bets in the same way. Place bet win probabilities, payoff odds, and house edges are as follows.

. . . .

Place Bet	Win Probability	Payoff Odds	House Edge
4	1/3	9 to 5	6.7%
5	2/5	7 to 5	4.0%
6	5/11	7 to 6	1.5%
8	5/11	7 to 6	1.5%
9	2/5	7 to 5	4.0%
10	1/3	9 to 5	6.7%

. . . .

As you can see from the table, house edges for place bets vary. Place bets on 6 or 8 have the lowest house edge: 1.5%. If you must make a place bet, make it on 6 or 8. The house edge for this bet is computed as follows.

$6 PLACE BET ON 6 (or 8)

. . . .

Win	Probability
7	5/11
−6	6/11

. . . .

$$E \text{ for } \$6 \text{ bet} = [7 \times (5/11)] - [6 \times (6/11)] = -1/11$$
$$E \text{ for } \$1 \text{ bet} = (-1/11)/6 = -1/66 = -.015$$
$$\text{House edge} = 1.5\%$$

Place Bets and Come-out Rolls. A place bet is "off" (ignored) on come-out rolls, unless you request otherwise, in which case an "on" marker is placed on your chips. The purpose of this rule is to confuse you.

BIG 6 AND BIG 8 BETS

You make a big 6 bet by placing chips in the **BIG 6** area in the corner of the betting layout (Figure 2). You make a big 8 bet by placing chips in the **BIG 8** area. You can do this any time. You win a big 6 bet if, starting with the next roll of the dice, 6 comes up before 7. You lose if 7 comes up before 6. You win a big 8 bet if 8 comes up before 7 and lose if 7 comes up before 8. The payoff odds are 1 to 1.

There are five ways to get 6 and six ways to get 7, for a total of eleven outcomes that determine the big 6 bet. As we have seen, $P(6 \text{ comes up before } 7) = 5/11 = P(\text{win big 6 bet})$. Ditto for a big 8 bet. The house edge is computed as follows.

$1 BET ON BIG 6 (OR BIG 8)

Win	Probability
1	5/11
−1	6/11

$$E = [1 \times (5/11)] - [1 \times (6/11)] = -1/11 = -.091$$
$$\text{House edge} = 9.1\%$$

Help! Big 6 and big 8 bets are exactly the same as place bets on 6 or 8, except your payoff odds are worse. The decrease in payoff odds, 1 to 1 instead of 7 to 6, causes the house edge to jump from 1.5% to 9.1%. Why would anyone in their right mind make a big 6 or big 8 bet instead of the equivalent place bet? To quote a Las Vegas gambler, "Every citizen is entitled to go broke however he wants."

HARDWAY BETS

A hardway bet is a bet that 4, 6, 8, or 10 comes up the "hard way" (the same number on each die) before 7 or some other way of getting that number comes up.

You win a hardway 4 bet if double-2 comes up before 7 or any other 4 combination (3 1 or 1 3). If 7, 3 1, or 1 3 comes up before double-2, you lose the hardway 4 bet. You win a hardway 6 bet if double-3 comes up before 7 or any other 6 combination (1 5, 5 1, 2 4, or 4 2). If 7, 1 5, 5 1, 2 4, or 4 2 comes up before double-3, you lose the hardway 6 bet.

You win a hardway 8 bet if double-4 comes up before 7 or any other 8 combination. You win a hardway 10 bet if double-5 comes up before 7 or any other 10 combination. You make a hardway bet by placing chips in one of the four hardway boxes in the center of the betting layout (Figure 2).

Suppose you make a hardway 6 bet. Since there are six combinations that yield 7 and four that yield 6 that aren't double-3, there are ten losing combinations and one winning combination. Thus, P(win hardway 6 bet) = 1/11. Similarly, P(win hardway 8 bet) = 1/11, P(win hardway 10 bet) = 1/9, and P(win hardway 4 bet) = 1/9.

Payoff odds are 7 to 1 for hardway bets on 4 and 10, and 9 to 1 for hardway bets on 6 and 8. Expected winnings and the house edge are computed as follows.

$1 BET ON HARDWAY 4 (OR 10)

Win	Probability
7	1/9
−1	8/9

$E = [7 \times (1/9)] - [1 \times (8/9)]$
$\quad = -1/9$
$\quad = -.111$

House edge = 11.1%

$1 BET ON HARDWAY 6 (OR 8)

Win	Probability
9	1/11
−1	10/11

$E = [9 \times (1/11)] - [1 \times (10/11)]$
$\quad = -1/11$
$\quad = -.091$

House edge = 9.1%

Hardway bets are great bets — for the house!

BUY BETS AND LAY BETS

Buy Bets. Buy bets are like odds bets and place bets: You bet that a point will come up before 7. Buy bets are like place bets: You don't have to make a buy bet in conjunction with another bet. Buy bets are like odds bets: You get fair payoff odds. Buy bets on 4 or 10 pay 2 to 1. Buy bets on 5 or 9 pay 3 to 2. Buy bets on 6 or 8 pay 6 to 5.

Here's the catch: When you make a buy bet, you have to pay the casino a 5% charge. For example, if you make a buy bet of $100, you must pay an additional $5 when you make the bet. This charge is kept by the casino whether you win or lose.

The 5% charge must be payable in multiples of the minimum table bet. For example, at a $1 minimum table, if you make a $40 buy bet, you pay a charge of 5% of $40, or $2. If you make a $10 buy bet, even though 5% of $10 is 50 cents, you still have to pay $1. If you feel you must make a buy bet, make it in multiples of 20 times the minimum bet.

To make a buy bet, you give chips to the dealer and specify which point you want to bet on. The dealer puts your chips in the appropriate point box with a "buy" marker on them so they can be distinguished from come bets that are put there after a point is established.

Since you get fair odds for buy bets, your expected winnings equals 0. However, since you pay a 5% charge, you are actually spending $1.05 and losing an average of $.05 (the charge) for every dollar bet. Your expected winnings per $1 bet (expected winnings divided by total bet) equals $-.05/1.05 = -.048$, yielding a 4.8% house edge. Bye bye, buy bets.

Lay Bets. Lay bets are the opposite of buy bets: You bet that 7 will come up before a specified point. You get fair odds. Lay bets on 4 or 10 pay 1 to 2: $P(7 \text{ before } 4) = 2/3$. Lay bets on 5 or 9 pay 2 to 3: $P(7 \text{ before } 5) = 3/5$. Lay bets on 6 or 8 pay 5 to 6: $P(7 \text{ before } 6) = 6/11$.

When you make a lay bet you have to pay the casino a 5% charge, but now the charge is 5% of the amount you would win, not the amount you bet. For example, if you make a lay bet of $200 on 4, and 7 comes up before 4, you win $100 (1 to 2 payoff odds). You pay the casino a charge of 5% of $100, or $5, to make a $200 lay bet on 4.

The house edge for lay bets depends on the point. Expected winnings for a fair odds bet are 0. For a lay bet on 4 or 10, you pay 5% of one-half of your bet (payoff odds are 1 to 2), which comes to $.025 per dollar bet. Thus, you actually spend $1.025 and lose an average of $.025 (the charge) per dollar bet. Your expected winnings per $1 bet equals $-.025/1.025 = -.024$, yielding a house edge of 2.4%.

The charge for a lay bet on 5 or 9 is 5% of two-thirds of your bet (payoff odds are 2 to 3), or $.033 per dollar bet. Expected winnings per dollar bet for a lay bet on 5 or 9 equals $-.033/1.033 = -.032$, yielding a house edge of 3.2%. The charge for a lay bet on 6 or 8 is 5% of five-sixths of your bet (payoff odds are 5 to 6), or \$.042 per dollar bet. Expected winnings per dollar bet for a lay bet on 6 or 8 equals $-.042/1.042 = -.040$, yielding a house edge of 4.0%.

THE BOTTOM LINE

The pass line, don't pass, come, or don't come bets, backed up with odds bets whenever possible, give the house an edge of less than 1%. This makes it possible for you to have an exciting evening of gambling before you go broke. Here is a table of craps bets and house edges.

CRAPS BETS AND HOUSE EDGES
(Assuming Proper Payoffs)

Bet	House Edge, %
Pass line	1.4
Come	1.4
Don't pass	1.4
Don't come	1.4
Above with odds bet	<1
Place bet on 6 or 8	1.5
Place bet on 5 or 9	4.0
Place bet on 4 or 10	6.7
Field bet	5.6
Any craps (2, 3, or 12)	11.1
Big 6 or big 8	9.1
Hardway 6 or 8	9.1
Hardway 4 or 10	11.1
One number (2, 3, 7, 11, 12)	16.7
Buy bets	4.8
Lay bet on 4 or 10	2.4
Lay bet on 5 or 9	3.2
Lay bet on 6 or 8	4.0
Craps–eleven bet	12–15

Final comment. Craps is a fast game. Even though other games have worse house edges, you'll probably lose just as much per hour playing craps.

CHAPTER

4

KENO AND SLOTS

I've got $500 and I'm going to lose it if it takes all night.

—DETERMINED GAMBLER

Keno is a game everyone can enjoy, from the manic gambler trying to find patterns in random numbers to the depressive conventioneer slumped over the bar. For a mere $2 per bet, you can sit in the keno lounge, sipping free drinks, dreaming big bucks.

Keno is played as follows. You mark from 1 to 12 numbers or "spots" on a keno card, which you turn in at a keno betting window. If you're not near the betting area, you give your ticket to a keno "runner," who will register your ticket for you. When your ticket is registered, you get a receipt showing the numbers you bet on.

A transparent box contains balls numbered from 1 through 80. The balls are blown around the box by air current. Twenty winning balls are blown at random into a chute. "At random" means that every ball has the same chance of being selected. You win if enough of the numbers you bet on are among the twenty winning numbers.

The most popular choice is a 10-number bet. Here is the payoff table for a $2, 10-number bet at Harrah's Casino in Lake Tahoe, Nevada.

PAYOFFS FOR $2, 10-NUMBER BET

Match	Pays
5	$ 4
6	40
7	280
8	1,800
9	8,000
10	50,000

47

For example, if 5 of your 10 numbers are among the 20 winning numbers, you are paid $4 (a $2 profit, since you paid $2 for the ticket). If 8 of your 10 numbers are among the 20 winning numbers, you are paid $1800. If all 10 of your numbers are among the 20 winning numbers, you are paid $50,000.

There are good reasons for the popularity of keno. First, it's easy to play. Keno runners circulate throughout the casino, allowing you to make bets from almost anywhere. Second, you can win a large amount with a small bet. Finally, keno games are played once every 5 to 15 minutes. This allows you to take your time going broke.

The chances of winning a big payoff are small. The chance of matching 10 of 10 is 1 in 10 million. The house edge in keno is terrible. It ranges from 25% to 50%. On the other hand, games with more favorable house edges, like craps, don't offer huge payoffs for a small bet. If you think it's fun to risk $2 for the chance of winning $50,000, by all means play. If you don't think it's fun, don't play. What's the point of losing money if you don't enjoy it?

COUNTING

In order to develop a formula for computing keno probabilities, we must venture into the world of combinatorics (counting).

Permutations. A "permutation" is an ordering of things. The things can be numbers, letters, books on a shelf, people in a line, songs on a record, whatever. For example, take the letters A B. There are two permutations: AB, BA.

How many permutations are there of the letters A B C? Any one of the three letters can start an ordering. Once a first letter is selected, either of the two remaining letters can follow it. Once the first two letters are selected, there is one letter left. There are thus $3 \times 2 \times 1 = 6$ permutations: ABC, ACB, BAC, BCA, CAB, CBA.

What about four letters: A B C D? The same reasoning applies. There are four starting letters, three letters that can follow each starting letter, two letters available for each starting sequence of two letters, and one letter left to end a sequence. This gives a total of $4 \times 3 \times 2 \times 1 = 24$ permutations:

ABCD, ABDC, ACBD, ACDB, ADBC, ADCB, BACD, BADC,
BCAD, BCDA, BDAC, BDCA, CABD, CADB, CBAD, CBDA,
CDAB, CDBA, DABC, DACB, DBAC, DBCA, DCAB, DCBA

A pattern emerges. For the five letters A B C D E, there are 5 ×
4 × 3 × 2 × 1 = 120 permutations. There are 6 × 5 × 4 × 3 × 2 ×
1 = 720 permutations of six things. There are 7 × 6 × 5 × 4 × 3 × 2 ×
1 = 5040 permutations of seven things, and so on. Trust me.

There is a convenient way to write the permutations formula. A
whole number followed by an exclamation mark is called "factorial" and
denotes that number multiplied by every number less than it, down to 1.
For example, 6!, read as "6 factorial," equals 6 × 5 × 4 × 3 × 2 × 1 =
720. Likewise, 7! = 7 × 6 × 5 × 4 × 3 × 2 × 1 = 5040.

> There are $n!$ permutations of n objects, where $n! = n \times$
> $(n - 1) \times (n - 2) \times \ldots \times 1$. By definition, $0! = 1$. (There's
> only one way to arrange nothing.)

There are 8! = 40,320 permutations of the words in the question,
"Is it not yet time for my painkiller?" There are 9! = 362,880 orderings
of nine people in a line. There are 10! = 3,628,800 orderings of ten
plants on a shelf. There are 11! = 39,916,800 permutations of eleven
different Elvis Presley songs.

There are 4! = 24 permutations of the letters A B C D. Now suppose
the letters are A A C D. There are still 24 permutations, but every time
you exchange an A with an A, you get the same arrangement. Since there
are two arrangements of transposed A's for each of the 24 permutations,
there are a total of 4!/2! = 24/2 = 12 *distinguishable* arrangements.

> AACD, AADC, ACAD, ADAC, ACDA, ADCA, CAAD,
> DAAC, CADA, DACA, CDAA, DCAA

How many distinguishable arrangements are there of the five letters
A A A B C? There are 5! = 120 permutations of five things. For each
permutation of the letters A A A B C, there are 3! = 6 ways to permute A's
among themselves. Thus, there are 5!/3! = 120/6 = 20 distinguishable
arrangements of the letters A A A B C.

How many distinguishable arrangements are there of the nine letters
A A A A B B B C D? There are 9! = 362,880 permutations of nine letters.
For each permutation, there are 4! ways to permute A's among themselves and 3! ways to permute B's among themselves, for a total of 4! ×
3! = 144 indistinguishable arrangements. This gives 9!/(4! × 3!) =
362,880/144 = 2520 distinguishable arrangements.

How many distinguishable arrangements are there of the ten letters
Y Y Y N N N N N N N? There are 10! = 3,628,800 permutations of 10 letters. For each permutation, there are 3! ways to permute Y's among

themselves and 7! ways to permute N's among themselves, giving a total of $3! \times 7! = 30{,}240$ indistinguishable arrangements of each of the 10! permutations. Thus, there are $10!/(3! \times 7!) = 120$ distinguishable arrangements of the letters Y Y Y N N N N N N N.

Combinations.

How many different committees of 3 people can be selected from 10 people? Pretend the 10 people are in a line. Denote each person selected to be on the committee by Y ("yes") and each person not selected by N ("no"). For example, selecting the first, third, and fifth people in the line to be on the committee is denoted by Y N Y N-Y N N N N N. Selecting the first, second, and third people is denoted by Y Y Y N N N N N N N. Thus, the number of committees of three people that can be selected from ten people is the same as the number of distinguishable arrangements of a list of three Y's and seven N's, which we have seen to be $10!/(3! \times 7!) = 120$.

How many 5-card poker hands can be dealt from an ordinary deck of 52 cards? Denote a selected card by Y and a card not selected by N. The number of ways to deal 5 cards from 52 is the same as the number of distinguishable arrangements of a list of 5 Y's and 47 N's, namely, $52!/(47! \times 5!) = 2{,}598{,}960$.

How many ways are there to select 6 packages of Hostess Ding Dongs from 20? Denote a selected package by Y and an unselected package by N. The number of ways to select 6 packages of Ding Dongs is the same as the number of orderings of 6 Y's and 14 N's: $20!/(6! \times 14!) = 38{,}760$.

These examples lead to the selections formula for combinations.

> There are $n!/k!(n-k)!$ ways to select k things from n. The formula $n!/k!(n-k)!$ is read as "n choose k" and for convenience is denoted by $\binom{n}{k}$. Sometimes, $\binom{n}{k}$ is called "combinations of n things taken k at a time."

Combinations, Probability, and Ding Dongs.

We continue with the Ding Dongs example. Suppose that 5 of the 20 packages contain prizes. You randomly select 6 packages. Find the chance that exactly 2 of the selected packages contain prizes.

There are $\binom{20}{6} = 38{,}760$ ways to select 6 packages of Ding Dongs from 20. There are 5 packages that contain prizes, so there are $\binom{5}{2} = 10$ ways to select 2 packages with prizes from 5. If 2 of the 6 selected packages have prizes, the other 4 don't have prizes. Since 5 of the 20 packages have prizes, the other 15 don't. There are $\binom{15}{4} = 1365$ ways to select 4 packages without prizes from 15. This gives a total of $\binom{5}{2}\binom{15}{4} = 13{,}650$ ways to select 6 packages of Ding Dongs, with 2 that have prizes and 4 that don't. Thus,

$$P(\text{two prizes}) = \frac{\binom{5}{2}\binom{15}{4}}{\binom{20}{6}} = .3522$$

What is the chance that 3 of the 6 selected packages contain prizes? There are $\binom{5}{3} = 10$ ways to select 3 packages with prizes from 5, and $\binom{15}{3} = 455$ ways to select 3 packages without prizes from 15, giving a total of $\binom{5}{3}\binom{15}{3}$ ways of selecting 6 packages of Ding Dongs, 3 with prizes and 3 without. The chance of this occurring equals

$$P(\text{three prizes}) = \frac{\binom{5}{3}\binom{15}{3}}{\binom{20}{6}} = .1174$$

What is the chance that all 6 of the selected packages have prizes? The chance is zero, since there are only 5 packages with prizes.

KENO PROBABILITIES

We can use the selections formula to compute keno probabilities. Suppose you make a bet on 10 numbers. What is the chance that 5 of your 10 numbers are among the 20 winning numbers?

There are 80 numbers altogether, 10 picked by you and 70 not picked by you. There are $\binom{80}{20}$ ways to select 20 winning numbers. There are $\binom{10}{5}$ ways for 5 of your 10 numbers to be among the 20 selected and $\binom{70}{15}$ ways for the other 15 selected numbers to be among the 70 not picked by you. Thus,

$$P(\text{match 5}) = \frac{\binom{10}{5}\binom{70}{15}}{\binom{80}{20}} = .0514277$$

Using the same method, we find

$$P(\text{match } 6) \quad = \frac{\binom{10}{6}\binom{70}{14}}{\binom{80}{20}} = .0114794$$

$$P(\text{match } 7) \quad = \frac{\binom{10}{7}\binom{70}{13}}{\binom{80}{20}} = .0016111$$

$$P(\text{match } 8) \quad = \frac{\binom{10}{8}\binom{70}{12}}{\binom{80}{20}} = .0001354$$

$$P(\text{match } 9) \quad = \frac{\binom{10}{9}\binom{70}{11}}{\binom{80}{20}} = .0000061$$

$$P(\text{match all } 10) = \frac{\binom{10}{10}\binom{70}{10}}{\binom{80}{20}} = .0000001$$

Here are the payoffs and probabilities for the $2, 10-number keno bet at Harrah's.

PAYOFFS AND PROBABILITIES FOR $2, 10-NUMBER BET
. . . .

Match	Pays	Probability
5	$ 4	.0514277
6	40	.0114794
7	280	.0016111
8	1,800	.0001354
9	8,000	.0000061
10	50,000	.0000001

. . . .

The Real Truth. To compute expected payoff for a keno bet, multiply the payoffs by their probabilities and add. Expected payoff for the $2, 10-number bet at Harrah's equals $1.41. Since you paid $2 for the ticket, to get expected winnings, subtract $2 from expected payoff: E(winnings) = $1.41 − $2.00 = −$.59. To find the house edge, divide by the total bet, drop the minus sign, and convert to a percentage:

$$\text{House edge} = 29.5\%$$

Bleak. However, there are other factors you might want to consider. Maybe you want to play the game that will give you the lowest hourly loss. This is the casino version of getting the highest hourly wage. Relax in the keno lounge. Make a $2 bet. Drink a free toast to the suckers at the craps table who are playing a better game than you but are somehow losing more money.

SLOT MACHINES

Slot machines are a casino mainstay. Many of today's slot machines are computerized versions of their mechanical forerunners. Slot machines are easy to use, don't require human interaction with the bettor, can't be beaten by skill, and require a minimum of space.

Most slot machines take quarters or dollar tokens. Some casinos have nickel and dime machines. You put one or more coins of the correct denomination in the machine and pull the handle. There are usually three or four "reels" of symbols (fruits, money signs, genies, weird icons, etc.) that spin, perhaps independently. There is a window on the front of the machine that shows the results. You win if certain configurations of symbols appear on the "center line" or indicated area in the window. For example, if all symbols in the center line are the same fruit, you win something.

Payoff odds are displayed on the front of the machine. When you win, coins pour into a bin. Lights flash. Bells, whistles, and sirens sound. You scoop up your winnings like a plundering pirate. Everyone is impressed. "Some people have all the luck," envious observers mutter. You strut your stuff and put another coin in the machine. Nobody pays attention as your booty dwindles to zero.

On New Year's Eve at the Las Vegas Hilton, there was a cordoned-off section of $500 slot machines. To play, you had to purchase $500 tokens from the casino. The casino was packed. There was action at the

$500 machines. At 11:50 p.m., casino employees handed out horns, hats, and streamers. At midnight, everyone stopped gambling and started singing and yelling "Happy New Year." The place went wild. At 10 past midnight, everyone was gambling again.

Slot Probabilities.

Here is how to compute slot machine payoff probabilities, assuming that the reels spin independently of each other and are equally likely to stop at any point (this assumption isn't always true).

Suppose a machine has three reels, with each reel having ten symbols. There are $10 \times 10 \times 10 = 1000$ three-symbol combinations that can appear on the display. Suppose the first wheel has one cherry and nine other symbols, the second wheel has three cherries and seven other symbols, and the third wheel has five cherries and five other symbols. There are thus $1 \times 3 \times 5 = 15$ three-cherry combinations. The probability of getting three cherries on this machine equals $15/1000 = .015$.

To determine the house edge, you must multiply all payoff amounts by their probabilities and add. House edges for slot machines generally range between 2% and 30%. Some casinos advertise "slots with 98% payback." Such a machine pays off 98 cents per dollar bet, giving the casino a 2% edge. If you feed $10 per minute into a 98% payback slot machine, you will lose about $12 per hour.

Slot machine payoff probabilities can be altered by changing the symbols on the reels. Some machines pay off more than others. The law of averages assures that if you play long enough you will lose. At the same time, the law of very large numbers assures that in a casino packed with slot machine players, there will be a few big winners. When you look around a casino, you may wonder how there can be so many winners when you're always a loser. Actually, most of these winners, like yourself, eventually become losers.

VIDEO POKER

One of the latest manifestations of slot machine technology is the immensely popular video poker. For one to six quarters, you play a hand of draw poker. Usually, a pair of jacks or better yields a payoff. The payoffs are frequent but small. You can play this game for a long time before going broke. Other gambling games, like keno, blackjack, and horse racing, have video versions, but video poker is the most popular.

SPORTS BETTING

A gambler was hopelessly hooked on football betting. Nothing else interested him. Unfortunately, he lost almost every bet he made. Finally, even his bookie felt sorry for him. "You lose all your bets," said the bookie. "Why don't you bet on hockey instead of football?"

"Hockey?" said the gambler in dismay. "But I don't know anything about hockey!"

Although illegal outside Nevada, sports betting is America's most popular gambling activity. According to a February 1990 article in the *Wall Street Journal*, more than $25 billion is bet on sports annually: $1.3 billion legally in Nevada and $26.3 billion illegally.

Sports betting is different from games like roulette and craps, because sports events aren't played under identical conditions and because odds are based on the oddsmaker's opinion of how the public will bet. Sports betting is similar to games like roulette and craps, because the casino makes a profit.

. . . .

Sports betting is a combination of luck and skill. My losses are due to bad luck and my wins are due to skill.

—VETERAN LAS VEGAS GAMBLER

SPORTS BOOKS
AND BOOKIES

A sports book is an establishment in Nevada where you can legally bet on a sporting event. A "bookie" is someone who will take your illegal bet on a sporting event. A few years ago, there were numerous independent sports books in Las Vegas and Reno that offered sports betting and wagering on horse and dog races. These places were the hangouts of cigar-chomping, Runyonesque characters who liked to talk sports, swill booze, and read the racing form. Then casinos discovered that sports books were good for business.

Caesar's Palace, the Las Vegas Hilton, and the Mirage casinos spent millions building computerized sports books with satellite dishes, beaming in games on giant screens in cavernous, high-tech auditoriums. Today, there are more than 40 sports books in Las Vegas casinos. Most of the independents have closed.

FOOTBALL BETTING

The Point Spread. In a football bet, the supposedly inferior team, or underdog, is given a handicap, called a "point spread." Synonyms for point spread are "spread," "line," "number," and "price."

In the 1990 Superbowl, the San Francisco 49ers were 11-point favorites over the Denver Broncos. For 49ers bettors to win, the 49ers had to win the game by more than 11 points. For Broncos bettors to win, the Broncos had to either win the game or lose by less than 11 points. If the 49ers won by exactly 11 points, a tie would result, and bets would be returned.

If a team wins after the point spread has been included, they are said to "cover the spread," or "cover." I have a friend who says that covering the point spread is "winning in reality."

The 49ers won the Superbowl, 55 to 10. The difference in scores was more than enough for the 49ers to cover the point spread. 49ers bettors won.

Here are examples from the 1989 football season.

The Detroit Lions played the Minnesota Vikings at Minnesota. Minnesota was a 12-point favorite. This was listed on the betting board at the sports books as

Lions

Vikings   12

The home team is listed second. The point spread is next to the favorite. The Vikings won the game, 24 to 17. The difference in scores is less than the point spread, so the Vikings didn't cover the spread. The Vikings won the game, but Vikings bettors lost. Lions bettors won.

The Atlanta Falcons played the New Orleans Saints at New Orleans. New Orleans was a 7-point favorite.

Falcons

Saints   7

The Saints won, 20 to 13. The point spread is equal to the difference in scores. The game was a tie for the bettors. Bets were returned.

The Los Angeles Raiders played the San Diego Chargers at San Diego. San Diego was a 2½-point underdog.

Raiders   2½

Chargers

Since the visiting team Raiders are the favorite, the point spread is listed next to their name. The underdog Chargers upset the Raiders, 14 to 12. The underdog won the game, covering the spread.

Note that when the spread is listed in half-point amounts, as in this example, there can't be a tie.

Payoff Odds and House Edge.

Payoff odds for point spread bets are 10 to 11. For every $11 bet, winners get $10 plus the amount of the wager. For example, a winning bet of $11 is paid $21, $10 in winnings plus the $11 bet. You need eleven wins for every ten losses to balance these odds ($10 \times 11 = 11 \times 10$). Eleven wins in twenty-one bets equals $11/21 = .524$. Thus, you must win more than 52.4% of your bets (assuming all bets are the same amount) to make a profit.

You bet on a football game. Having no special knowledge of football, you toss a coin to decide which team to bet on. The chance equals .5 that your team covers the point spread: $P(\text{your team covers spread}) = .5$. The 10 to 11 payoff odds give the sports book a 4.5% edge. This is computed in the usual way: Find expected winnings and divide by the total amount bet. For an $11 bet, we have the following:

$11 FOOTBALL BET

. . . .

Win	Probability
10	.5
−11	.5

. . . .

$$E = (10 \times .5) - (11 \times .5) = -.5$$
$$E(\text{for \$1 bet}) = -.5/11 = -.045$$
$$\text{House edge} = 4.5\%$$

Splitting the Action.

If the same amount is bet on each team in a game, the sports book makes a guaranteed 4.5% profit, no matter who wins. This is what the book wants: no risk. The book's profit is called "juice," "vigorish," or "vig."

For example, suppose five $11 bets are made on each team, for "action" of $110. No matter which team covers, the $50 in winning payoffs (10 to 11 odds) comes from the $55 in losing bets, leaving a $5 profit for the book. Expressed as a percentage of the action, the sports book makes a profit of 5/110 = 4.5%.

The Oddsmaker.

The person who sets point spreads is called the "oddsmaker" or "linemaker." There are only a few oddsmakers in Las Vegas. They work as consultants for the sports books. Bookies around the country use Las Vegas lines. The same oddsmakers who make football point spreads give odds for other sports. The leading oddsmaker in Las Vegas is Michael "Roxy" Roxborough. Roxy and his company, Las Vegas Sports Consultants, provide sports books with point spreads and odds for all major sports.

When setting point spreads, the oddsmaker considers the relative strengths and weaknesses of competing teams. In order to produce a point spread that evenly divides the betting action, the oddsmaker must also be a good judge of how the public will bet.

Point spreads have been accurate in many situations. In the 1984 through 1988 NFL seasons, the home team won 57.8% of games but covered the spread in 48.9% of games. Home underdogs won 37.4% of their games but covered the point spread in 51.8% of games.

In the 1969 Superbowl between the New York Jets and the Baltimore Colts, the Colts opened 17-point favorites. Heavy betting came in on the Colts, and the point spread moved up to 18. The Jets won, 17 to 6. After the game, the media blasted Las Vegas oddsmakers. How could the Colts have been made 18-point favorites? The reporters didn't understand the point spread system. Even though the Jets were a superior team, the bettors loved the Colts. The point spread split the action, guaranteeing the books their 4.5% vig. A lower point spread would have resulted in a bigger profit but would have put the books at risk.

The Opening Line.

After Sunday's pro football games have been played, the oddsmaker determines point spreads for the coming week. The first public point spreads for the week's games are called the "opening line." The Stardust Casino in Las Vegas is traditionally the first sports book to post the opening football lines, early Sunday evening. Wise guys (big bettors) wait at the betting windows with wallets full of money so they can take advantage of "mistakes" in the opening line before they are corrected. During the first hour of betting, the Stardust Sports Book has the energy level of the New York Stock Exchange.

Moving the Spread.
A bettor gets the point spread posted when a bet is made. If the betting action on a game isn't evenly divided, the sports book can lose if the overbet team covers the spread. In order to minimize this risk, the book will change the point spread to encourage betting on an underbet team.

Because of point spread changes, spreads may vary in different sports books. The variation is small, since wary bettors notice disparities and bet accordingly.

Suppose a sports book has a $5000 bet limit on a game. If a wise guy bets the limit, the book may move the line half a point to attract action on the other side. Big bettors often have other people, known as "beards," bet for them in order to bet more than the limit on a game. Some bettors use many beards. I know a wise guy named Mel who has a dozen or so beards who work for him. Mel's minimum bet on a football game is $50,000. A friend told me, "When Mel bets, E. F. Hutton listens."

Another source of action is money bet by bookies. Bookies outside Nevada don't have the same flexibility as sports books for moving point spreads. If action is unbalanced, a bookie can bet the overbet team with another bookie to balance his own books. This is called "laying off" bets. If the overbet team wins, the bookie wins his laid-off bet and doesn't suffer a large loss. Bookies with small operations lay off bets with bigger bookies, and so on, until the money finally reaches Nevada sports books. Bookies don't like to lay off bets, because they have to take the worst of the 10 to 11 odds. Some bookies prefer to risk losing.

THE PROS VERSUS
THE POINT SPREAD

Once after I did a radio show in Las Vegas, a sports bettor friend asked if I wanted to "go shopping" with him. It was after midnight, but people stay up late in Vegas. I assumed we'd go to an all-night grocery store. It turned out that my friend wasn't going grocery shopping, but shopping for point spreads. We went to different sports books in search of the best spreads. Bleary-eyed, we finished shopping at 2 a.m.

Because sports books want a point spread that evenly divides the action, if the betting public likes a bad team, a clever bettor may have an edge. It's easy to tell when a point spread evenly divides the action, but how do you know if the spread is distorted due to the public's misconceptions? Football handicappers spend many hours handicapping games during the pro football season, looking for good betting opportunities. My friend the Lone Wolf says that smart handicappers betting against a naive public are like "a shark against a crippled goldfish." However, if there are a lot of sharks and only a few goldfish, the system becomes unstable. At that point the big shark—the oddsmaker—steps in, moves the point spread, and gobbles everyone up.

Middles.
Sometimes, no matter what the oddsmaker does, point spread manipulation backfires. Sports books take a risk if they make a large point spread move to equalize betting action. If the point spread is changed and the difference in score lands between point spreads, the sports book gets "middled" and loses both ways.

In the 1979 Superbowl between the Pittsburgh Steelers and Dallas Cowboys, the line opened with the Steelers 4½-point favorites. Heavy betting came in on the Cowboys, and the point spread was lowered to 3½. Bettors who had been reluctant to give 4½ points decided 3½ was reasonable, and there was a flurry of betting on the Steelers. By game time, the line had moved to 4 to encourage more Cowboys bets.

The Steelers won, 35 to 31. Those who bet the Steelers as a 3½-point favorite won their bets, those who bet the Cowboys as a 4½-point underdog won their bets, and those who bet on either team with the Steelers a 4-point favorite tied. The sports books took a bath.

Some bettors shop for point spread differences. If they find a large difference, they bet on both teams and hope the difference in score lands between the spreads. If this happens, they win both bets. This is a middle for the bettor. In the 1979 Superbowl, if you had bet the Steelers as a 3½-point favorite and the Cowboys as a 4½-point underdog, you would have middled the sports books and won both bets.

When you bet on both teams at different point spreads, you can either win both bets, win one and lose one (for a loss of the vig), or win one and tie one. In order to make a profit betting middles, you have to win both sides more than 4.8% of the time (excluding ties).

The Plumber.
It was the end of football season. A friend of mine owed a local bookie a couple dollars (a gambling "dollar" is a hundred real dollars). She made the drop-off at a bar where the bookie hung out. On the phone, the bookie told her to put the money in an envelope, leave it with the bartender, and tell him it was for the "plumber." She did as instructed. The next time she talked to him, she asked why he was called the "plumber."

"Because I *am* a plumber," said the bookie. "I need an ordinary job so I can pay the rent during losing streaks."

OVER-UNDER BETS

An "over-under," or "totals," bet is a bet that the combined scores of both teams will be either "over" or "under" a specified number. It makes no difference who wins the game. The over-under number is listed in the sports book next to the underdog. For example, suppose the New York Jets are playing the Buffalo Bills:

Bills   43

Jets   2½

The home team Jets are 2½-point favorites. The over-under line is 43. If you bet "over" and the total score of both teams is over 43, you win. If the total score of both teams is under 43, you lose. If you bet "under" and the total score is less than 43, you win. If the total score is over 43, you lose. If the total score equals 43, both over bets and under bets tie.

Suppose the Jets win, 25 to 21. The total is 46. The over-under line is 43, so over bettors win, and under bettors lose. Suppose the Jets win, 14 to 7. The total is 21. Over bettors lose, and under bettors win. Suppose the Bills win, 28 to 15. The total is 43. Totals bets tie, and money is returned.

Payoff odds on over-under bets are 10 to 11, the same as point spread bets. The over-under line is supposed to represent the "median" total. That is, there should be a 50–50 chance that the total score will be over and a 50–50 chance it will be under. As with point spread wagers, the sports book's goal is simply to split the betting action on each side. If the betting is evenly divided on over and under, the sports book makes a guaranteed 4.5% profit, just as with point spread bets. Over-under lines are moved (like point spreads) to balance betting action.

A Smoking Wager. Lisa decided to quit smoking. I thought she wouldn't quit for long, but a mutual friend thought she would quit for good. It's difficult to bet that someone will quit smoking for good because if this happens, the bet will never be settled. I decided to make this an over-under bet. On the basis of Lisa's past record, I thought that there was a 50–50 chance that she would stop for about 17 days. I made the over-under number 17. I bet my friend that Lisa would smoke a cigarette in "under" 17 days, and he bet "over." We didn't tell Lisa, so she couldn't rig the bet for a piece of the action. She smoked a cigarette after dinner in 15 days, with the comment, "The main joy of eating is to have a cigarette afterward." Under bettors won.

PARLAY BETS

A parlay bet is a single bet on two or more games. For you to win a parlay bet, every team you pick has to cover the point spread. If any of your picks fails to cover, you lose the bet. You get the point spreads posted on the betting board at the time you make the bet.

The payoff odds for parlay bets depend on the number of games bet. In many sports books, you can make parlay bets on two, three, four, or five games.

Parlay payoff odds, win probabilities (assuming each team has a 50–50 chance of covering), and corresponding fair odds are as follows:

PARLAY BETS

· ·

No. of Teams	Payoff Odds	P(win)	Fair Odds
2	13 to 5	1/4	3 to 1
3	6 to 1	1/8	7 to 1
4	10 to 1	1/16	15 to 1
5	20 to 1	1/32	31 to 1

· ·

Since the only way to win a parlay bet is if all teams cover, the win probabilities in the table are obtained by using the multiplication rule for independent events, taking 1/2 to the appropriate power. For example,

$$P(\text{win two-team parlay}) = (1/2)^2 = 1/4$$
$$P(\text{win three-team parlay}) = (1/2)^3 = 1/8$$

Using these win probabilities and payoff odds, house edges for parlay bets are computed as follows:

TWO-TEAM PARLAY
· · · · · · · · · ·
Expected Winnings for $5 Bet

Win	Probability	
		$E = -.5$
13	1/4	E for $1 bet $= -.5/5 = -.10$
−5	3/4	House edge $= 10\%$

· · · · · · · · · · · · · ·

THREE-TEAM PARLAY
. . . .
Expected Winnings for $1 Bet

Win	Probability
6	1/8
-1	7/8

$E = -.125$

House edge $= 12.5\%$

. . . .

The house edge for a four-team parlay equals 31.3%. The house edge for a five-team parlay equals 34.4%. Assuming that P(any team covers) = .5 and games are independent, parlay bets have bad house edges for the player. Unless your win probability is considerably higher than .5 for at least one game, parlay bets are bad.

For ordinary parlay bets, if there is a tie, the wager is reduced to the next lowest parlay. For example, a point spread tie in a three-team parlay makes the wager a two-team parlay on the remaining picks.

Parlay Cards. In addition to ordinary parlay bets, most sports books have parlay cards, which are printed at the beginning of each week, usually with a combination of pro games and college games. The point spreads and over-under lines printed on the card apply to the parlay card bet. You can typically make from 3 to 10 picks, with the corresponding payoff odds printed on the card. In order to win a parlay card bet, all your picks must win.

Parlay cards are different in some respects from parlays made at the betting window. You can pick more games on parlay cards. With ordinary parlays, you get the currently posted line. With parlay cards, you get the point spread on the card. On most parlay cards, payoffs odds are expressed as "for 1" instead of "to 1." For example, on the Las Vegas Hilton SuperBook parlay card (Figure 6), the payoff for a three-game parlay is "3 for 3 pays 6 for 1." "6 for 1" means winners get back $6 for every $1 bet, including the dollar bet. Ordinary payoff odds "6 to 1" means winners get back $6 for every dollar bet, *plus* the dollar bet. To convert "for one" into "to one" odds, subtract a dollar. For example, "6 for 1" is the same as "5 to 1."

The effect of ties on parlay card bets varies and is stated on the card. For example, ties win on Hilton SuperBook parlay cards (Figure 6). On some parlay cards, all lines are listed in half-point amounts, so there are no ties.

Parlay Cards Outside Nevada. Illegal parlay card betting is popular in bars and other gambling hangouts. Payoff odds on illegal parlay cards are worse than payoff odds on legal Nevada cards. And if you win, there's

SuperBook_{SM}

PARLAY CARD

TIES WIN *TIES WIN*

$5.00 minimum   **$5.00** minimum

Pay Schedule	
3 for 3 pays	6 for 1
4 for 4 pays	11 for 1
5 for 5 pays	20 for 1
6 for 6 pays	40 for 1
7 for 7 pays	80 for 1
8 for 8 pays	150 for 1
9 for 9 pays	300 for 1
10 for 10 pays ..	600 for 1

1. Ties win.
2. Minimum bet $5.00. Maximum bet subjective to the decision of management.
3. The Las Vegas Hilton pays off only on the ticket (receipt) generated from the stub portion of this card. This card is not valid for payment.
4. Games not played on date specified or invalid selections are considered no action and reduce card to next lowest number of plays offered.
5. Tickets must have a minimum of three propositions in action or wager is void and money will be refunded.
6. Management reserves the right to refuse any wager(s) or delete or limit any selection(s) prior to the acceptance of any wagers.
7. Printed parlay card point spreads at the time the wager is accepted are used in the determination of winners, losers and ties.
8. All Las Vegas Hilton Sports Book House Wagering Rules and Regulations apply to parlay cards unless stipulated otherwise on card.
9. $150,000 maximum payoff on any card and all identical cards. This aggregate payoff on all identical cards shall be paid in proportion to the amount wagered.
10. Winning tickets are void after 60 days from date of last event. Winning tickets may be mailed in for redemption. Please use registered mail and enclose a self-addressed stamped envelope. Mail to Las Vegas Hilton, P.O. Box 15086, Las Vegas, Nevada 89114
11. Please check your ticket (receipt) for accuracy before leaving the wagering counter.
12. The Las Vegas Hilton is not responsible for lost, stolen, altered or unreadable tickets.
13. These cards are for the express use of the Las Vegas Hilton Sports Book and are not valid for any other use. We employ no agents.

$5.00 Super Progressive Parlay
14 for 14 pays .... Progressive Jackpot
$40,000.00 Minimum

— SUPER PROGRESSIVE PARLAY RULES —

1. The Super Progressive Parlay is any fourteen team parlay available on the Las Vegas Hilton, Flamingo Hilton and Reno Hilton Sports Book Parlay Cards.
2. The cost per fourteen team parlay is $5.00.
3. The progressive jackpot will be $40,000.00 minimum each week offered. The Las Vegas Hilton, Flamingo Hilton, and Reno Hilton will provide the initial $40,000.00.
4. Each week that the progressive jackpot is not awarded, all monies wagered on the fourteen team progressive parlay will be added to the following week's jackpot.
5. Each week's jackpot will be printed on the parlay card or publicly posted in the Sports Book.
6. The jackpot award is aggregate. If there are multiple winning tickets, the jackpot will be divided equally.
7. All Parlay Card Rules apply unless stipulated otherwise in Super Progressive Parlay Rules.

LAS VEGAS HILTON
Race & Sports SuperBook
P.O. Box 15086 • Las Vegas, Nevada 89114

FIGURE 6 (FRONT)

✕ SuperBook℠

TIES WIN PARLAY CARD

PROGRESSIVE JACKPOT: $41,200

★ DENOTES HOME TEAM

WK. #2

COLLEGE FOOTBALL - THURS., SEPT. 8, 1988

1 TEXAS	CTV	5:00 PM	2 B.Y.U. ★	+ 3
3 GAME TOTAL OVER 44			4 GAME TOTAL UNDER 44	

COLLEGE FOOTBALL - SAT., SEPT. 10, 1988

5 WASHINGTON		9:15 AM	6 PURDUE ★	+ 11
7 EAST CAROLINA		10:00 AM	8 VIRGINIA TECH ★	+ 2
9 GEORGIA ★		10:00 AM	10 T.C.U.	+ 15
11 BAYLOR		10:10 AM	12 KANSAS ★	+ 19
13 OHIO STATE ★		12:40 PM	14 SYRACUSE	+ 2
15 SOUTHERN CAL	TV	12:40 PM	16 STANFORD ★	+ 13
17 WYOMING		1:00 PM	18 LOUISVILLE ★	+ 9
19 OKLAHOMA	CTV	2:00 PM	20 NO. CAROLINA ★	+24
21 FLORIDA		4:00 PM	22 MISSISSIPPI @Jackson	+ 11
23 PENN STATE		4:00 PM	24 VIRGINIA ★	+ 2
25 INDIANA		4:00 PM	26 RICE ★	+ 15
27 AUBURN ★	CTV	9:30 AM	28 KENTUCKY	+ 16
29 ALABAMA		4:00 PM	30 TEMPLE ★	+ 15
31 FLORIDA STATE ★		4:00 PM	32 SOUTHERN MISS.	+32
33 TULANE		4:10 PM	34 IOWA STATE ★	+ 8
35 TENNESSEE ★		4:30 PM	36 DUKE	+ 13
37 NEBRASKA	TV	5:00 PM	38 U.C.L.A. ★	Pk
39 GAME TOTAL OVER 47			40 GAME TOTAL UNDER 47	
41 WASHINGTON ST.		5:00 PM	42 MINNESOTA ★	+ 2
43 VANDERBILT ★		5:00 PM	44 MISSISSIPPI ST.	+ 5
45 NOTRE DAME ★	TV	6:00 PM	46 MICHIGAN	+ 2
47 GAME TOTAL OVER 38			48 GAME TOTAL UNDER 38	
49 ARIZONA ★		6:30 PM	50 TEXAS TECH	+ 8
51 ARIZONA STATE ★		7:30 PM	52 ILLINOIS	+20

PRO FOOTBALL - SUN., SEPT. 11, 1988

53 GIANTS ★	TV	10:00 AM	54 49ERS	+ 2
55 GAME TOTAL OVER 43			56 GAME TOTAL UNDER 43	
57 BILLS ★		10:00 AM	58 DOLPHINS	+ 4
59 GAME TOTAL OVER 44			60 GAME TOTAL UNDER 44	
61 REDSKINS ★	TV	10:00 AM	62 STEELERS	+ 8
63 GAME TOTAL OVER 43			64 GAME TOTAL UNDER 43	
65 PACKERS ★		10:00 AM	66 BUCS	+ 6
67 GAME TOTAL OVER 42			68 GAME TOTAL UNDER 42	
69 BEARS		10:00 AM	70 COLTS ★	+ 2
71 GAME TOTAL OVER 41			72 GAME TOTAL UNDER 41	
73 SAINTS		10:00 AM	74 FALCONS ★	+ 8
75 GAME TOTAL OVER 44			76 GAME TOTAL UNDER 44	
77 RAMS ★		1:00 PM	78 LIONS	+10
79 GAME TOTAL OVER 43			80 GAME TOTAL UNDER 43	
81 BRONCOS ★		1:00 PM	82 CHARGERS	+13
83 GAME TOTAL OVER 44			84 GAME TOTAL UNDER 44	
85 BROWNS ★		1:00 PM	86 JETS	+ 9
87 GAME TOTAL OVER 40			88 GAME TOTAL UNDER 40	
89 EAGLES ★		1:00 PM	90 BENGALS	+ 5
91 GAME TOTAL OVER 47			92 GAME TOTAL UNDER 47	
93 SEAHAWKS ★		1:00 PM	94 CHIEFS	+ 8
95 GAME TOTAL OVER 44			96 GAME TOTAL UNDER 44	
97 OILERS ★	TV	1:00 PM	98 RAIDERS	+ 2
99 GAME TOTAL OVER 39			100 GAME TOTAL UNDER 39	
101 VIKINGS ★		1:00 PM	102 PATRIOTS	+ 6½
103 GAME TOTAL OVER 45			104 GAME TOTAL UNDER 45	

COLLEGE FOOTBALL - SUN., SEPT. 11, 1988

105 AIR FORCE	CTV	6:30 PM	106 SAN DIEGO ST. ★	+ 8
107 GAME TOTAL OVER 46			108 GAME TOTAL UNDER 46	

PRO FOOTBALL - MON., SEPT. 12, 1988

109 CARDS ★	TV	5:00 PM	110 COWBOYS	+ 2
111 GAME TOTAL OVER 45			112 GAME TOTAL UNDER 45	

MARK SELECTED TEAMS, NUMBER OF TEAMS, AND AMOUNT BET WITH AN ●

● MUST REMAIN INSIDE THE CIRCLE.

FIGURE 6 (BACK)

always the chance your bookie will be arrested before you get paid. Numerous states are considering legalizing parlay card betting. As of this writing, Oregon is the only state to have done so.

TEASER CARDS

Teaser bets are parlays where point spreads are adjusted in the bettor's favor. Payoff odds are lower than ordinary parlay payoffs. On the Caesar's Palace teaser card in Figure 7, you can make four to ten picks that are "teased" by 6 points.

Compare the Caesar's teaser card to the Hilton parlay card in Figure 6. In the first NFL game on the Hilton card, the 49ers are 2-point underdogs against the Giants. On the teaser card, there are two numbers, Giants +4 and 49ers +8. This is what you get by adding 6 points to the point spread for each team. (The 49ers as 2-point underdogs plus 6 makes them 8-point underdogs. The Giants as 2-point favorites plus 6 makes them 4-point "underdogs"). If you pick this teased game, you get the indicated number of points for whichever team you pick. You can pick both teams. For example, if you bet on both teams in the 49ers–Giants game, if the Giants lose by less than 4 points or win by less than 8 points, you win both picks.

Compare payoff odds. On the teaser card, if you win four picks out of four you get paid 7 for 2 (3½ for 1), while on the parlay card, four for four pays 11 for 1. On the teaser card, five for five pays 5 for 1, while on the parlay card, five for five pays 20 for 1. On the teaser card, ten for ten pays 25 for 1, while on the parlay card, ten for ten pays 600 for 1. You make a substantial sacrifice in odds to get teased point spreads.

You win teaser bets if all your teams cover the adjusted point spread. If the actual difference in scores is close to the point spread for all games in your teaser bet, you'll win, no matter which teams you pick. In order for teaser cards (or anything else) to be a good bet, the fraction of bets that you win has to outweigh the payoff odds. For example, if you make a four-team teaser bet with 5 to 2 (7 for 2) payoff odds, you have to win more than $2/(5+2) = 2/7 = 28.6\%$ of your bets in order to make a profit.

According to oddsmaker Michael Roxborough, in the 1985–1986 season, NFL home teams covered point spreads teased by 6 points 70.8% of the time. Assuming independence of games, this makes the chance of winning a four-game, randomly selected, home-team teaser equal $.708^4 = .2513$, less than the 28.6% long-run win percentage needed to beat the 5 to 2 payoff odds. In this situation, the house edge for a four-team, 6-point teaser card bet with 5 to 2 payoff odds equals 12%.

MONEY LINE BETS

A "money line" bet is a wager in which odds are used without point spreads. The bet is determined by who wins the game. Sports books offer money line bets on selected games.

A money line bet consists of two payoff odds, one for the favorite and one for the underdog. Odds for the favorite are preceded by a minus sign. In most cases odds for the underdog are preceded by a plus sign. Money line payoff odds are posted relative to a $100 bet. For example, suppose the Washington Redskins are playing the New York Giants and the money line is

Redskins   +160

Giants   −180

The team listed second is the home team. If you bet on the favorite Giants, you "give" the odds and bet $180 to win $100 (the payoff odds are 1 to 1.8). For example, suppose you bet $18 on the Giants. If they win the game, you win $10. If you bet on the Redskins, you "take" the odds and bet $100 to win $160 (the payoff odds are 1.6 to 1). If you bet $10 on the Redskins and they win the game, you win $16. No point spread.

In money line betting, there is an optimal betting allocation for the sports books analogous to the evenly split action for point spread bets. The sports book wants the betting action to be roughly proportional to the odds.

If the book gave "fair" odds, then odds for the favorite would be the reverse of the odds for the underdog. The money line for the favorite would be the negative of the money line for the underdog. For example, if the odds on the favorite were 1 to 4, fair odds on the underdog would be 4 to 1. The money line would be −400 for the favorite and +400 for the underdog. If the betting action were proportional to the odds, four-fifths of the action would be on the favorite and one-fifth on the underdog. Bets on the loser would pay off bets on the winner, and the book would break even. For example, if $400 were bet on the favorite and $100 on the underdog, and the underdog won, the $400 in losing favorite bets would pay the 4 to 1 odds due underdog bettors. If the favorite won, the $100 in losing underdog bets would pay the 1 to 4 payoff odds due favorite bettors.

The book wants to make a profit, not break even, so odds on the underdog are lower than the reverse of odds on the favorite. If the oddsmaker is doing a good job, betting action will match odds, and the book will make a profit no matter who wins. If the oddsmaker is doing

CAESARS PALACE
• TEASER •
PARLAY CARD
- TEAMS MUST WIN - TIES LOSE -

• TEASER PRICES •	
4 for 4 Pays ...... 7 for 2	7 for 7 Pays ..... 10 for 1
5 for 5 Pays ...... 5 for 1	8 for 8 Pays ..... 14 for 1
6 for 6 Pays ...... 7 for 1	9 for 9 Pays ..... 18 for 1
10 for 10 Pays..... 25 for 1	

TEASER RULES

1. Teams must win — ties lose.
2. $2.00 minimum bet and $500 maximum bet.
3. This card is valid only as a means of recording selections; the official parlay card wager will be paid consistent with the selections shown on the betting ticket. Player is responsible for verifying that selections on the parlay card match the selections on the official betting ticket. A player's parlay card stub is not valid for payment. Betting tickets are issued only at Caesars Palace Olympiad Sports Book.
4. Games must be played on the date specified, unless otherwise designated by management. Invalid selections and any game chosen that is postponed or suspended shall be deemed "NO ACTION" and the wager refunded.
5. Tickets must have a minimum of four propositions in action, or wager is void and money will be refunded.
6. Management reserves the right to refuse any wager(s) or delete or limit any selection(s) prior to the acceptance of wagers.
7. Printed point spreads at the time the wager is accepted are used in the determination of winners, losers and ties.
8. Where applicable, all Caesars Palace Olympiad Sports Book posted rules and regulations apply to this card, unless otherwise stated on the card.
9. The total amount paid out on identical winning wagers shall not exceed $150,000. If the total amount to be paid out on identical winning wagers would otherwise exceed $150,000, winners will be paid in proportion to the amounts wagered.
10. Betting ticket void 60 days from date of last event on card.
11. Players must be 21 or older.
12. There is no limit as to how many times a person may place a wager each week.
13. Wagers must be placed prior to the start of the first selected event on the card.
14. Winning tickets are paid immediately after the official results of the last selected game are posted.
15. Caesars Palace will pay applicable Federal Wagering Taxes.
16. Caesars Palace allows no agents. Each person must place his/her own wager.

MARK SELECTED TEAMS, NUMBER OF TEAMS AND AMOUNT WAGERED BY BLACKING OUT ENTIRE NUMBER WHILE REMAINING WITHIN BOX.

FIGURE 7 (FRONT)

CAESARS PALACE
LAS VEGAS

COLLEGE FOOTBALL • THU., SEP. 8, 1988

| 1 Texas | +2 | 5:00p | 2 B. Y. U. | TV | +10 |

COLLEGE FOOTBALL • SAT., SEP. 10, 1988

3 Washington	-5	9:15a	4 PURDUE		+17
5 East Carolina	+4	10:00a	6 VIRGINIA TECH		+8
7 GEORGIA	-9	10:00a	8 t. c. u.		+21
9 MICHIGAN STATE	-12	10:00a	10 Rutgers		+24
11 Baylor	-12	10:10a	12 KANSAS		+24
13 BOSTON COLL.	-13	10:30a	14 Cincinnati		+25
15 Iowa	-26	11:00a	16 KANSAS STATE		+38
17 ARKANSAS	-13	12:00p	18 Tulsa		+25
19 COLORADO STATE	+2	12:00p	20 Hawaii		+10
21 OHIO STATE	+4	12:40p	22 Syracuse	TV	+8
23 Southern Cal	-7	12:40p	24 STANFORD	TV	+19
25 Wyoming	-2	1:00p	26 LOUISVILLE		+14
27 Oklahoma	-18	2:00p	28 NO. CAROLINA	TV	+30

AT JACKSON

29 Florida	-4	4:00p	30 MISSISSIPPI		+16
31 Penn State	+4	4:00p	32 VIRGINIA		+8
33 Indiana	-8	4:00p	34 RICE		+20
35 AUBURN	-10	9:30a	36 Kentucky		+22
37 Alabama	-9	4:00p	38 TEMPLE		+21
39 FLORIDA STATE	-27	4:00p	40 Southern Miss		+39
41 Tulane	-2	4:10p	42 IOWA STATE		+14
43 TENNESSEE	-7	4:30p	44 Duke		+19
45 Nebraska	+5	5:00p	46 U. C. L. A.	TV	+7
47 Washington St.	+5	5:00p	48 MINNESOTA		+7
49 VANDERBILT	EV	5:00p	50 Mississippi St.		+12
51 NOTRE DAME	+4	6:00p	52 Michigan	TV	+8
53 ARIZONA	-3	6:30p	54 Texas Tech		+15
55 ARIZONA STATE	-14	7:30p	56 Illinois		+26

NFL FOOTBALL • SUN., SEP. 11, 1988

57 GIANTS	+4	10:00a	58 49ers	TV	+8
59 BILLS	+2	10:00a	60 Dolphins		+10
61 REDSKINS	-2	10:00a	62 Steelers	TV	+14
63 PACKERS	-1	10:00a	64 Buccaneers		+13
65 Bears	+4	10:00a	66 COLTS		+8
67 Saints	-2	10:00a	68 FALCONS		+14
69 RAMS	-4	1:00p	70 Lions		+16
71 BRONCOS	-7	1:00p	72 Chargers		+19
73 BROWNS	-3	1:00p	74 Jets		+15
75 EAGLES	+1	1:00p	76 Bengals		+11
77 SEAHAWKS	-1	1:00p	78 Chiefs		+13
79 OILERS	+4	1:00p	80 Raiders	TV	+8
81 VIKINGS	EV	1:00p	82 Patriots		+12

COLLEGE FOOTBALL • SUN., SEP. 11, 1988

| 83 Air Force | -2 | 6:30p | 84 SAN DIEGO STATE | | +14 |

NFL FOOTBALL • MON., SEP. 12, 1988

| 85 CARDINALS | +4 | 5:00p | 86 Cowboys | TV | +8 |

FIGURE 7 (BACK)

a terrible job, everyone will bet on the same team. If this team wins, the sports book goes under.

Money Line Differences.
In the Redskins–Giants example (Redskins +160, Giants −180), the absolute difference between payoff odds on the favorite and payoff odds on the underdog is 20. This is called a "20-cent" line. Redskins +160, Giants −190, is a "30-cent" line. Redskins +160, Giants −200, is a "40-cent" line. With optimal betting allocation, the larger the odds disparity, the more profit there is for the book. As the odds get higher, however, the edge diminishes. In a game where one team is a big favorite, the difference between favorite and underdog odds has to be larger than for an evenly matched game to assure the book a profit.

Even Games.
If teams are evenly matched, the game is called a "pick" game. For pick games, money line odds are negative for both teams. For example, the money line

49ers   −110

Bears   −110

means that the payoff odds are 10 to 11 for a bet on either team. This is the same as if the point spread were 0 in a point spread bet. The money line

Colts   −120

Chargers   Even

means that if you bet on the favorite Colts, the payoff odds are 10 to 12, and if you bet on the underdog Chargers, the payoff odds are even, or 1 to 1.

Money Lines and Point Spreads.
There is a correlation between point spreads and money line odds: The greater the point spread, the greater the money line odds. Sports books don't like to offer money lines for games with point spreads larger than 8 because the chance is too great that the favorite will win the game. Statistician Hal Stern used data from the 1981, 1983, and 1984 NFL seasons to estimate the chance that a team favored by a given number of points will win the game. With this model, money line odds can be estimated from point spreads. The odds in Stern's model are somewhat similar to the point spread to money line conversions used by oddsmaker Michael Roxborough. The following table has probabilities estimated by Stern's model and incorporates both Roxy's and Stern's money line odds.

Point Spread	P(favorite wins)	Typical Money Line Odds
1	.53	−115, −105
2	.56	−130, +110
3	.58	−150, +130
3.5	.60	−160, +140
4	.61	−170, +150
5	.64	−190, +160
6	.67	−220, +180
7	.70	−260, +220

FUTURES BETS

A futures bet is a bet on who will win a future event, such as the Super-bowl or a conference championship. Futures bets have payoff odds for each team or proposition.

For example, on January 29, 1990, the day after the 1990 Super-bowl, the "Stardust Hotel Race and Sports Book" posted odds on the winner of the 1991 Superbowl. The odds, listed in the February 2 *Stardust Sports Registry*, and subject to change according to betting patterns, are shown in Figure 8. The San Francisco 49ers are favorites at 7 to 1. Next are the Denver Broncos at 8 to 1. You could get 20 to 1 on the Kansas City Chiefs, 45 to 1 on the Seattle Seahawks, 80 to 1 odds on the Atlanta Falcons, and 125 to 1 on the lowly Dallas Cowboys. By February 10, enough had been bet on the 49ers to lower their odds to 3 to 1.

Payoff odds on futures bets are set to give the sports book a substantial edge when the betting action is roughly proportional to the odds.

WEIRD BETS

For some games, bets are offered that aren't normally available. For example, most books offer unusual bets on the Superbowl, the most heavily bet game of the year.

In the 1990 Superbowl between the San Francisco 49ers and the Denver Broncos, the following bets were established by oddsmaker Michael Roxborough.

STARDUST

COMPLETE SPORTS SCHEDULE

STARDUST HOTEL RACE AND SPORTS BOOK
VOL. 1 - ISSUE 9 - FEB. 2 - FEB. 22, 1990

ODDS TO WIN 1991
SUPER BOWL

San Francisco 49ers	7
Denver Broncos	8
LA Rams	8
NY Giants	10
Minnesota Vikings	10
Philadelphia Eagles	12
Cleveland Browns	15
New Orleans Saints	15
Buffalo Bills	18
Cincinnati Bengals	18
Green Bay Packers	18
Kansas City Chiefs	20
Houston Oilers	20
Washington Redskins	20
LA Raiders	25
Pittsburgh Steelers	25
Chicago Bears	25
Miami Dolphins	25
San Diego Chargers	30
Detroit Lions	35
Indianapolis Colts	40
New England Patriots	45
Seattle Seahawks	45
Tampa Bay Buccaneers	60
Atlanta Falcons	80
NY Jets	100
Phoenix Cardinals	125
Dallas Cowboys	125

Odds posted January 29, 1990
No bets accepted prior to Jan. 29
Odds subject to change without warning.

FIGURE 8

Player to Score First Touchdown

Jerry Rice	7 to 2
John Taylor	4 to 1
Roger Craig	5 to 1
Van Johnson	6 to 1
Ricky Nattiel	8 to 1

49ers Score No Points   40 to 1
Broncos Score No Points   15 to 1

Number of Jerry Rice Catches

Over 5½	−170
Under 5½	+130

There Is a Score in Last Two Minutes of First Half

Yes	−220
No	+180

Shortest Field Goal

49ers	−150
Broncos	+110

Total Rushing and Receiving Yards by Roger Craig

Over 117	−120
Under 117	−120

49ers Win by the Score 7–0   200 to 1
49ers Win by the Score 5–0   5000 to 1

There Is a Safety

Yes	+600
No	−900

The Game Goes into Overtime

Yes	+800
No	−1200

Opening Coin Toss

Heads	−110
Tails	−110

Professional gamblers regard bets like these as sucker bets. If a sports book gets careless, however, such a bet can be advantageous to the bettor. For example, in Superbowl XX between the Chicago Bears and New England Patriots, a few sports books posted the proposition that William "The Refrigerator" Perry would score a touchdown. It seemed unlikely that Perry, a defensive lineman used occasionally on offense during the regular season, would have a chance to score in the Superbowl. Initially, you could get 12 to 1 odds if you bet Perry would score a touchdown. This proposition was bet so heavily that by game time the odds had been reduced to 2 to 1. You couldn't bet that Perry wouldn't score a touchdown. Perry scored a touchdown.

PREDICTABILITY

. . . .

Leopards break into the temple and drink to the dregs what is in the sacrificial pitchers; this is repeated over and over again; finally it can be calculated in advance, and it becomes part of the ceremony.

When the surf flings a drop of water on to the land, that does not interfere with the eternal rolling of the sea, on the contrary, it is caused by it.

——FRANZ KAFKA, *PARABLES AND PARADOXES*

When is the occurrence of an event due to chance, and when can it be predicted in advance? Suppose the event is the result of a football game. Football isn't like craps or roulette. There are no theoretical probabilities.

Sampling. The purpose of a sample is to provide a representative picture of a population or process. Football bettors look at samples of past games to decide how to bet. Can past games be used to predict the outcomes of future games? This is a difficult question. A basic problem in statistics is to distinguish between a predictable and chance pattern.

For example, many believe that teams play better at home than on the road. The home team gets psyched up for Monday night football and the national TV audience. If the home team is an underdog, there might even be an upset. Are Monday night home underdogs a good bet?

During the 1979 through 1989 regular seasons, Monday night home underdogs covered the point spread in 41 out of 58 games, for a point spread win–loss record of 70.7%. Was this a random fluke or a prime betting opportunity?

Here is an important principle of statistical reasoning:

If an observed event is highly unlikely to occur under a given hypothesis and less unlikely under a plausible alternative hypothesis, reject the given hypothesis.

Let's apply this principle to Monday night home underdogs. The observed event is 41 point spread wins in 58 games. If Monday night home underdogs had a 50–50 chance of covering (our given hypothesis), in 58 games they would cover the spread about half the time, or in 29 games. It turns out that 41 or more point spread wins in 58 games is extremely unlikely if each team has a 50–50 chance of covering (the probability is about .001). Thus, we reject the 50–50 hypothesis and conclude that Monday night home underdogs have a greater than 50–50 chance of covering the spread.

You should take advantage of this investment opportunity while it lasts. The heavy betting on Monday night home underdogs resulting from big bettors reading this book will cause oddsmakers to adjust the point spreads, and the pattern will vanish like a mirage.

Which Measurements Should You Use?

. . . .

It was an extra-ordinarily bitter day, I remember, zero by the thermometer. But considering it was Christmas Eve there was nothing extra-ordinary about that. . . . Seasonable weather for once, in a way. . . .

It was a glorious bright day, I remember, fifty by the heliometer, but already the sun was sinking down into the . . . down among the dead. . . .

It was a howling wild day, I remember, a hundred by the anenometer. The wind was tearing up the dead pines and sweeping them away. . . .

It was an exceedingly dry day, I remember, zero by the hygrometer. Ideal weather, for my lumbago.

—SAMUEL BECKETT, *ENDGAME*

Handicappers spend hours studying data, trying to discover patterns that will enable them to predict winners. For example, a team's performance in its last game, a team's next opponent, winning streaks, and injury information are considered important factors. Computer Sports World, in Boulder City, Nevada, gives computer users access to a wide variety of sports data. Mort Olshan's "Gold Sheet," in Los Angeles, California, is a sports information newsletter with accurate and detailed statistics. The question is this: Which factors (if any) can be effectively used to predict point spread winners?

Don't ask me. Oddsmakers take much into account when setting the point spread. The point spread structure is sensitive to predictable patterns, because as soon as a pattern is discovered, bettors make big bets and the lines are adjusted. An injury to a key player may affect the outcome of a game, but this information will be incorporated into the spread. Weather information is incorporated into the spread as well.

. . . .

Las Vegas oddsmaker Michael Roxborough keeps up-to-date, accurate weather information. A Chicago gambler once called Roxborough from Chicago on a Sunday morning. "Roxy," said the gambler, "I'm thinking of going to the Bears game this afternoon but only if the weather's nice. I have a hangover and don't feel like looking out the window. Could you tell me what the weather's like here in Chicago?"

MONEY MANAGEMENT AND THE KELLY SYSTEM

What Is Your Real Chance of Winning? Suppose you spend a lot of time compiling and analyzing sports data and you use this data in making your picks. Maybe you can do better than tossing a coin to decide how to bet. Maybe not.

Keep a record of your wins and losses. Use your actual win percentage as your chance of winning. Remember, because of the 10 to 11 odds, you have to win more than 52.4% of your bets to make a profit. If you win less than that, no system will help you.

Suppose you can pick point spread winners with 60% accuracy: $P(\text{your team covers spread}) = .6$. You compute expected winnings in the usual way:

EXPECTED WINNINGS FOR AN $11 BET

. . . .

Win	Probability
10	.6
−11	.4

. . . .

$$E = 10 \times .6 + (-11) \times .4 = 1.6$$
$$E \text{ for } \$1 \text{ bet} = 1.6/11 = .145$$

In repeated play, you'll win an average of 14.5 cents for every dollar bet. You have a 14.5% edge, almost three times better than the 5.3% house edge in roulette.

Say you start the season with a bankroll of $1100. If you bet it all on the first bet, your expected winnings are $160. This is great, but if you lose, you're broke. Even though you have an edge, there's a 40% chance you'll lose. Your chance of lasting through 10 bets with a .60 win percentage, betting your entire bankroll each time, equals $.60^{10} = .006$. Even in a favorable situation, you should only bet a fraction of your bankroll. How much?

Fixed Fraction Betting. In 1956, J. L. Kelly Jr. published an article about mathematical information theory applied to transmission of information over a phone line. The results were applied to gambling and became known as the Kelly system. Using the Kelly system, you always bet a fixed fraction of your bankroll (the total amount you have for betting purposes). More precisely, you bet the fraction of your bankroll that maximizes its "growth rate." Warning: The next sentence is technical. Growth rate is the expected logarithm of return, where return is payoff per dollar bet. Here's an informal explanation.

What does it mean to bet a fixed fraction of your bankroll, for example, 10%? It means that for each bet, you bet 10% of what you have left. Suppose you have $1100 to bet on football games. Your first bet would be 10% of $1100, or $110. If you win, the 10 to 11 payoff odds yield $100 in winnings, so you have $1200. Your second bet would be 10% of $1200, or $120, and so on. On the other hand, if you lose your first bet of $110, you would have $1100 − $110 = $990. On your second bet, you bet 10% of $990, or $99. Whenever you bet, you bet 10% of your current bankroll. Contrary to certain popular (and silly) gambling strategies in which you bet big to cover your losses, with the Kelly system the more you have, the more you bet, and the less you have, the less you bet.

Statistician Leo Breiman showed that the Kelly system is optimal for two reasons. First, it will do better in the long run than any substantially different strategy. Second, the expected number of bets necessary to reach a specified goal with the Kelly system is lower than with any other strategy.

A third desirable property of the Kelly system is that since you always bet a fixed fraction of your bankroll, it's difficult to go broke. In fact, if money were "infinitely divisible" (it's not), you would always be able to bet a fraction of what you had left and would never go broke.

Warning: The Kelly system applies *only* to favorable bets. A
bet is favorable if the player has positive expected winnings.
If the house has an edge, the Kelly system says take up tennis.

A football bet with 10 to 11 odds is favorable if the chance of
winning is greater than .524. When making a single bet, if p is your win
probability, the Kelly system says bet the fraction $2.1p - 1.1$ of your
current bankroll. For example, if your chance of winning is .55, bet the
fraction $2.1 \times .55 - 1.1 = .055$, or 5.5% of your bankroll. If you have
$1000, bet $55. If your chance of winning is .60, bet the fraction $2.1 \times
.6 - 1.1 = .16$ of your bankroll. If your chance of winning is .70, bet 37%
of your bankroll.

The Kelly system can be used for simultaneous bets, that is, betting
on more than one game at the same time. The following table gives the
percentage of your bankroll you should bet on each game when betting on
one to five games, when your chance of winning is .55, .60, or .70. For
example, if you bet on four games at the same time and your chance of
winning each bet is .60, you should bet 15% of your bankroll on each
game, for a combined bet of 60% of your bankroll.

PERCENTAGE OF BANKROLL TO BET ON
EACH GAME WITH KELLY SYSTEM

. . . .

	CHANCE OF WINNING		
No. of Games Bet	.55	.60	.70
1	5.5%	16%	37%
2	5%	16%	33%
3	5%	15%	28%
4	5%	15%	24%
5	5%	14%	19%

. . . .

How fast does your bankroll grow if you use the Kelly system? If
your chance of winning is .55 and you bet 5.5% of your bankroll on each
bet, in a sequence of 100 bets your bankroll will increase by about 15%.
For example, $1000 will grow to $1150. If your win probability is 0.60
and you bet 16% of your bankroll on each bet, in 100 bets your bankroll
will be multiplied by a factor of about 3.23. If your win probability is .65
and you bet 26% of your bankroll on each bet, in 100 bets $1000 will
grow to more than $25,000.

SPORTS SERVICES

Sports services are companies that sell picks and information, supposedly to help you become a winning bettor. For example, you might get "key picks" for each week's games by calling a 900 number. The increased popularity of football betting has created a cottage industry of sports services and touts. There are more than 700 sports services nationwide, with many listed in a publication called the "Sports Betting Guide."

There are so many sports services that some services will have winning records just due to chance (the law of very large numbers). Sports services usually have different categories of picks. For example, a service may have "key picks," "early picks," "late picks," "five star picks," "gold label picks," and "technical picks." Each category of picks favors different sets of games. If the "key picks" record is 65% winners, the sports service advertises "Key picks — 65% winners!" without mentioning the categories with losing records.

Another ploy is to charge customers only for winning picks. "You pay only for winners!" claims the ad. Suppose you must pay the service $100 for winning picks and nothing for losing picks. Suppose the service is tossing coins to decide how to pick. They'll pick winners about half the time. You pay $100 for winning picks and nothing for losing picks, an average of $50 per pick. Meanwhile, you'll be betting. You may not have to pay the sports service for losing picks, but you'll be paying your bookie.

I recently saw an ad in which a sports service claimed it could "run rings around Wall Street" and pick 80% winners. If you pick 80% winners, the Kelly system says bet 58% of your bankroll on each bet. In 100 bets, you would win an average of 80 bets and lose 20. If you started with $1000, you would end up with $15 billion. If you think these guys really pick 80% winners, I've got a bridge I'd like to sell you.

LEGALIZATION

The Oregon Lottery. Oregon has legalized football betting in its lottery. Oregon lottery football bets are made with parlay cards. Unlike Nevada sports book parlay cards, which have fixed payoff odds, Oregon lottery parlay cards have pari-mutuel payoffs: Winning bettors split the prize pool after the state takes out a percentage. In Oregon, the percentage is 50%, the same as the California Lotto game, and more than the 17% track take in horse racing, another pari-mutuel wagering system.

You can bet on three to ten games. If all your picks are winners, you split the prize pool with other winners who picked the same number of games. For example, if you pick five games and all five of the teams you select cover the point spread, you split the prize pool with everyone else who went five for five. In this case, the prize pool equals the total amount of money bet on five game picks after the state has taken out 50%.

Should Sports Betting Be Legal?

The National Football League protested when Oregon introduced sports betting. I think they protested too much. The NFL owes much of its commercial success to football betting and the media coverage it receives. Almost every major newspaper in the country has daily information for sports bettors (Figures 9 and 10). TV sponsors know that viewers who bet on a televised game are as captive as an audience can get.

If betting on football is legal in Nevada and Oregon, why should it be illegal elsewhere? Billions are bet illegally. Why not give states a cut of the action.

What about fixed games? Underworld connections? Oddsmaker Michael Roxborough points out that sports books and the NFL have the

San Francisco Chronicle

SCOREBOARD

WEEKEND SPORTS

TODAY

BASKETBALL—College Men: Cable Car Classic at Santa Clara (consolation game, 6 p.m.; championship game 8:30 p.m.); Cal St. Los Angeles vs. San Francisco St., 7:30 p.m.

BASKETBALL—College Women: Cardinal Classic at Stanford (consolation game, 6 p.m.; championship game 8 p.m.); UOP vs. St. Mary's, 5:15 p.m.; Fresno Pacific vs Hayward St., 7:30 p.m.

HORSE RACING—Thoroughbreds, Bay Meadows, San Mateo, post time 1 p.m.; inter-track betting, Golden Gate Fields, Albany, first post 1 p.m.

TOMORROW

HORSE RACING—Thoroughbreds, Bay Meadows, San Mateo, post time 1 p.m.; inter-track betting, Golden Gate Fields, Albany, first post 1 p.m.

TRANSACTIONS

BASEBALL

Senior Professional Baseball Association WINTER HAVEN SUPER SOX—Fired Ed Nottle, manager. Named Leon Roberts manager.

BASKETBALL

National Basketball Association DENVER NUGGETS—Signed forward Mike Higgins to a 10-day contract.
Continental Basketball Association CEDAR RAPIDS SILVER BULLETS — Signed Reggie Owens, forward. Released David Colbert, forward.

FOOTBALL

National Football League
LOS ANGELES RAMS—Activated James Washington, safety, from injured reserve. Placed Clifford Hicks, cornerback, on injured reserve.

HOCKEY

National Hockey League
CALGARY FLAMES—Called up goalie Steve Guenette from the Salt Lake Golden Eagles of the International Hockey League.
NEW JERSEY DEVILS—Assigned Walt Poddubny, center, to Utica of the American Hockey League for conditioning purposes.

COLLEGE

DARTMOUTH—Named Bob Whalen head baseball coach.
ELON COLLEGE—Named Mike Hardin head baseball coach.
FORT HAYS STATE—Named Bob Cortese head football coach.
HUNTER—Announced the resignation of Gary Wohlstetter, head volleyball coach and assistant men's basketball coach.
TOLEDO—Named Kent Baker head track and field and cross-country coach.

BETTING EDGE

NFL PLAYOFFS

TOMORROW

Favorite	Spread	Underdog
PHILA.	3	L.A. Rams
HOUSTON	6½	Pittsburgh

COLLEGE FOOTBALL

TODAY

JOHN HANCOCK BOWL
In El Paso, Texas

Favorite	Spread	Underdog
Texas A&M	6	Pittsburgh

FREEDOM BOWL
In Anaheim

Washington	3	Florida

PEACH BOWL
In Atlanta

GEORGIA	2	Syracuse

GATOR BOWL
In Jacksonville, Fla.

Clemson	7½	West Virginia

TOMORROW
COPPER BOWL
In Tucson, Ariz.

ARIZONA	6½	N.C. State

MONDAY
HALL OF FAME BOWL
In Tampa, Fla.

Auburn	8	Ohio State

CITRUS BOWL
In Orlando, Fla.

Illinois	3	Virginia

COTTON BOWL
In Dallas

Arkansas	1	Tennessee

FIESTA BOWL
In Tempe, Ariz.

Florida State	4	Nebraska

ROSE BOWL
In Pasadena

USC	2	Michigan

ORANGE BOWL
In Miami

Notre Dame	2	Colorado

SUGAR BOWL
In New Orleans

Miami-Fla.	9	Alabama

COLLEGE BASKETBALL

TODAY

Favorite	Spread	Underdog
NOTRE DAME	15	Butler
Louisville	10½	KENTUCKY
MICHIGAN	20½	E Michigan
UNLV	4	Arkansas
GEO WASH	4½	Geo Mason
XAVIER	10	Bowling Green
OLD DOM.	3½	Jas Madison
DAYTON	11½	Bradley
TCU	16	E Carolina
SMU	14	Tulane
ST LOUIS	10	Indiana St

NBA

TODAY

Favorite	Spread	Underdog
NEW YORK	16	Orlando
UTAH	14½	Miami
WASH.	2	Chicago
CLEVELAND	3	Phoenix
DETROIT	15½	New Jersey
INDIANA	3	Atlanta
Houston	3½	CHARLOTTE
DALLAS	3½	Denver
MILWAUKEE	13	Minnesota
LA CLIPPERS	1	Philadelphia

Home team in capitals. NL: No line.

Spreads from the official Las Vegas line

FIGURE 9

BETTING EDGE/RAY RATTO

If nothing else, you must say this for the National Football League. It packs its bad games in a tidy, one-day package for you.

For instance, neither game on Sunday, the Rams-Eagles nor the Steelers-Oilers, should be viewed without the benefit of special, triple-thick welder's glasses, to minimize exposure to untrained or underdeveloped retinas.

But because they have been bestowed with the grandiose title "wild-card game," we all know we're going to bet on them. We are ashamed of ourselves, but we're hooked, and we freely admit it.

Thus, opinions are demanded, and we are nothing if not accommodating.

■ L.A. Rams plus 3 at Philadelphia: The Rams are, by all measures, the superior team, which means it ought to be a simple matter of taking the right side and queueing up at the bank window. Of course, there is a nagging suspicion that the Rams will wake up Sunday morning, see the mercury dancing at 15 above with a wind-chill factor of 800-below, say, "The hell with it," and go home.

Still, when you look at the Eagles, you should see a team with Randall Cunningham, a great defensive line and little else. This is just the spot for the Rams to show that they are indeed the second-best team in the conference, perhaps in the NFL, and getting a field goal to boot. Too easy to pass up.

■ Houston minus 6½ at home against Pittsburgh: Now this one really smells like Puget Sound on a humid day. The Oilers no more deserve to give this many points than does Rice University, because with few exceptions they have hearts of balsa and haven't won an important game in what seems like an epoch. In addition, Jerry Glanville is as unappealing a character in his way as, say, Bobby (The Brain) Heenan is in his and is less amusing to boot. House Of Pain indeed.

The problem, though, with playing the Steelers is that first-time playoff teams usually find their hands at their throats by halftime. Remember all the noise people made two years ago about the Saints in their first appearance? Remember how they lost to the Vikings by about a thousand points? Besides, other than Louis Lipps, what's so special about the Steelers? A quarterback named Bubby? Sorry, but this one's a big-time no-sell.

A cautionary postscript, though. If you must play this game, bet no more than the change in your pocket, unless you are a bus driver and accustomed to carrying $33 worth of quarters at any given time. Neither one of these teams deserves our attention, let alone our available cash. Go out, breathe the brisk air, duck those household chores, leave the kids with unsuspecting neighbors, and above all, avoid viewing this holiday turkey. It'll be the best four hours you never spent.

■ LAST WEEK: 3-1, with victories by New Orleans, Kansas City and Detroit, and a loss on those liver-hearted Oilers.

■ THIS SPACE THIS SEASON: 40-38, including 10-3 the last three weeks. God, but we're good.

■ THIS SPACE SINCE 1986: 145-135-3.

NFL ODDS

NFL PLAYOFFS			HOUSTON	6½	Pittsburgh
SUNDAY			Home team in capitals. NL: No line.		
Favorite	Spread	Underdog			
PHILADELPHIA	3	L.A. Rams	Spreads from the official Las Vegas line		

FIGURE 10

common interest of maintaining the integrity of the game. If games are fixed, books lose money, and the betting public loses interest. Do people bet on professional wrestling? Anyway, problems exist, whether or not betting is legal.

Sports betting is a recreational activity enjoyed by people from all walks of life. You can bet in the stock market on the success (or failure) of a company. Why shouldn't you be able to bet on the success of a football team without risking a jail term? Attempts to manipulate odds or fix a football game should be treated the same as insider trading.

BASKETBALL, BOXING, BASEBALL, AND HOCKEY

Basketball. Basketball betting is like football betting. There is a point spread to equalize the betting action, and you get 10 to 11 odds for a bet on either the favorite or underdog. There are over-under bets, parlay bets, teaser bets, money line bets, and parlay card bets. In fact, most things about basketball betting are like football betting, even the devotion of bettors.

· ·

Harry was a basketball betting nut from New York. A few years ago, in the middle of basketball season, Harry's Uncle Irv died. Harry was upset by his uncle's death, but there was an important basketball game on the day of the funeral. Harry decided to go to the funeral and wear his Walkman so he could listen to the game. As the casket was being lowered into the grave, Harry burst into tears. One of Harry's cousins came over to console him. "This was a real blow. You and Irv were pals," said the cousin. "Pals, shmals," sobbed Harry. "The Celtics just beat the Knicks by seven, and I had a grand on the Knicks plus six."

Boxing. There is no point spread in boxing. Betting is done with money lines. The money line structure is the same as football money lines. For example, suppose Smith is fighting Jones. Smith is the favorite.

Smith	−400
Jones	+300

San Francisco Chronicle   Monday, February 12, 1990

Gambler KO'd By $100,000 Bet on Tyson

Associated Press

Las Vegas

There's no such thing as a sure thing, and one unlucky gambler found that out the hard way when he plunked down $100,000 on Mike Tyson to beat Buster Douglas.

Before the fight Saturday night, Tyson was a 35-to-1 favorite, the biggest odds ever posted on a title fight.

"This was like Secretariat running against a Clydesdale," said Russ Culver, sports book manager at The Mirage. "This wasn't suppos-ed to be a real title fight. It was only supposed to be an exhibition."

At The Mirage, the only legal sports book to even post odds on the fight because of Tyson's utter dominance of the heavyweight division, betting started with Tyson a record 35-to-1 favorite.

But even that lopsided line was bet up by gamblers who continued to put money on the undisputed and undefeated champion. By the time the opening bell rang, Tyson was a 42-to-1 favorite.

Odds on Tyson fights have steadily gone up with each bout, and peaked previously at 31-to-1 when he knocked out Carl (The Truth) Williams in the first round of their fight last year.

Jimmy Vaccaro, the hotel's sports book director, said one gambler bet $1,500 on Douglas at 38-to-1 odds, making off with about $57,000.

FIGURE 11

This money line indicates that bettors on the favorite Smith must wager $400 to win $100, and bettors on the underdog Jones must wager $100 to win $300. The outcome of a money line bet is determined by who wins the fight. You may also be able to bet whether a boxer will win by a knockout.

For many fights, you can also bet how long the fight will last. Such bets are similar to over-under bets in football and basketball. For example,

Smith – Jones Fight

Does go 6½	−120
Does not go 6½	Even

In this case, if you bet $120 that the fight goes over 6½ rounds and it does, you win $100. (The payoff odds are 10 to 12, equivalently, 5 to 6.) You get even money odds if you bet that the fight goes under 6½ rounds. Since a round lasts 3 minutes, a half round equals 90 seconds. If the fight ends with a knockout exactly 90 seconds into the sixth round, the wager is a "push" (tie) and bets are returned.

Baseball.

Like boxing bets, baseball bets are made with money lines, not point spreads. You can also make over-under bets. A typical wager is posted on the betting board like this:

Rangers	Ryan	+180	7	Over	−120
A's	Stewart	−220	7	Under	Even

The Rangers are the visiting team. The A's are the home team. Ryan is the scheduled starting pitcher for the Rangers, and Stewart is the scheduled starting pitcher for the A's. The Rangers are the underdog, with payoff odds 1.8 to 1. If you bet $100 on the Rangers and the Rangers win the game, you win $180. (You get back $280, $180 in winnings plus the $100 you bet.) The A's are the favorite, with payoff odds 1 to 2.2 If you bet $220 on the A's and the A's win the game, you win $100. (You get back $320, $100 in winnings plus the $220 you bet.)

Baseball bets have a pitcher option. You can have your bet "on" only if the scheduled pitchers start the game, or you can have your bet apply no matter who pitches, with possible adjusted odds if one or both of the scheduled pitchers doesn't start.

Baseball over-unders are like football over-unders. You win an over bet if the total number of runs scored by both teams is more than the posted number, lose if the total is less than the posted number, and tie if the total is the same as the posted number. You win an under bet if the number of total runs scored is less than the posted number, lose if the total is more than the posted number, and tie if the total is the same as the posted number. In the A's–Royals example, the posted number is 7. If the A's win, 5 to 3, over bets win and under bets lose. If the Royals win, 3 to 0, under bets win and over bets lose. If the A's win 4 to 3, over-under bets tie.

The payoff odds for baseball over-unders are not always 10 to 11, as they are in football. In the A's–Rangers example, over bettors must bet $120 to win $100 (odds are 1 to 1.2). Under bettors get even payoff odds.

As with football and basketball, you can make baseball parlay bets and futures bets on who will win the championship. Some books offer "run lines," which are like point spreads.

Hockey. Hockey betting incorporates point spread, money line, parlay, and over-under bets. A typical wager looks like this.

Detroit	+1½	+120
Quebec	−1½	−140

Quebec is a 1½-goal favorite. This is equivalent to a point spread of 1½ in a football or basketball game, except instead of 10 to 11 payoff odds for bets on either team, the payoff odds are determined by the money line. In this case, if you bet on the favorite, Quebec, you have to bet $140 to win $100 (payoff odds of 1 to 1.4). If you bet on the underdog, Detroit, you bet $100 to win $120 (payoff odds of 1.2 to 1). You win a bet on Quebec if Quebec wins the game by two or more goals. You win a bet on Detroit if Detroit wins, ties, or loses by one goal.

A "split line" has half a goal difference between the favorite line and the underdog line. For example:

Detroit	+1½	Even
Quebec	−2	Even

In this wager, the payoff odds are the same (even) for a bet on either team, but the spreads are different. You win a bet on Quebec if Quebec wins the game by more than two goals. You win a bet on Detroit if Detroit wins, ties, or loses by one goal. In either case, with evenly split betting, the books break even. If Quebec wins by exactly two goals, bets on Quebec tie, and bets on Detroit lose, so the book wins all money bet on Detroit. According to oddsmaker Michael Roxborough, 17% of hockey games during the period 1983 to 1987 landed on the split line. Given that percentage, if the betting is evenly split and action is the same on each game, the vig will be half of 17%, or 8.5%.

THE BOTTOM LINE

If you're going to gamble, which games should you play? I recommend sports betting. There are two reasons for this:

1 – Unlike in roulette, craps, and keno, it's impossible to prove that you can't win in the long run.

2 – When you win, it's because you're smart, and when you lose, it's because somebody fumbled.

CHAPTER

....

6

....

BLACKJACK

Grain upon grain, one by one, and one day, suddenly, there's a heap, a little heap, the impossible heap.

—SAMUEL BECKETT, *ENDGAME*

Blackjack is the only casino game where the player can have an edge over the casino. Winning blackjack strategies were developed by computer simulations in the 1950s and publicized in *Beat the Dealer*, by mathematician Ed Thorp. Thorp's book caused thousands of gamblers to flock to the casinos.

Fortunately for the casinos, winning strategies were difficult and gave the player only a slight edge under ideal conditions. Multiple-deck games were introduced. Most blackjack players lost. Pros were cheated or banned from the tables.

Suppose you master a winning blackjack strategy. You will have perhaps a 2% edge. How much you win in the long run depends on how much you bet. If you bet $5 on each hand and play one hand per minute, you'll bet $300 per hour. With a 2% edge, your hourly winnings will average 2% of $300, or $6. After travel expenses, your earnings will be below minimum wage. Even though you're a winner, you'd make more money working at Burger King.

In this chapter, I'll show you how to play blackjack well enough to break even.

HOW TO PLAY

Blackjack, or "21," is a card game played between the dealer (a casino employee) and one or more players. The object is to get a score higher than the dealer's score without exceeding 21, the best possible score.

8 7
.

Your score is the sum of the values of each card in your hand. Cards 2 through 9 are worth face value. Tens, jacks, queens, and kings are worth 10, and aces can be either 1 or 11.

The game is played at a table with a betting layout, as in Figure 12. Before any cards are dealt, you bet by placing chips in the circle in front of you on the betting layout. After bets are made, the dealer deals two cards to everyone. One of the dealer's cards is dealt face up ("up card"), and the other face down ("hole card"). A player's first two cards are usually dealt face down.

After two cards are dealt to everyone, play begins with the first player to the left of the dealer. There are two main choices: "stand," take no additional cards, and "draw" (or "hit"), take an additional card. To stand, slide your cards under your chips (Figure 13). To draw, scratch your cards toward yourself on the table. If you draw, the dealer deals you a card, face up. After looking at your new card, you again decide whether to stand or draw, and so on. When you finally stand, your hand is complete, and play moves to the player on your left.

If your score ever exceeds 21, you "bust" and must turn your cards face up. If you bust, you lose.

When the bettors are finished, if any haven't busted, the dealer plays her hand. Unlike the players, the dealer stands or draws according to fixed rules. The dealer must draw with a score less than 17 and stand with a score of 17 or more.

If the dealer busts, any players who haven't busted win. If the dealer doesn't bust, players with a score higher than the dealer win and those

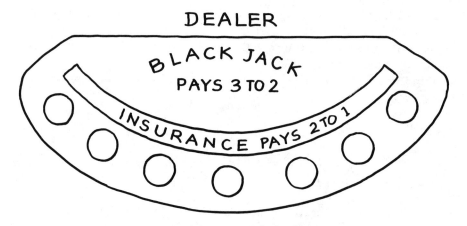

FIGURE 12

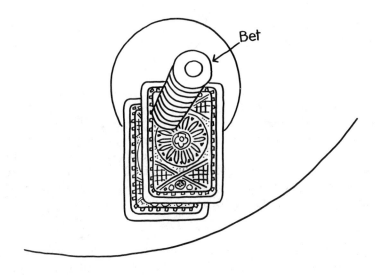

FIGURE 13

with a lower score lose. Players with the same score as the dealer tie. Because players go first, a bettor who busts loses even if the dealer busts later. Payoff odds on regular hands are even, or 1 to 1.

Soft and Hard Hands.
If a hand contains an ace that can be counted as 11 points without the score exceeding 21, it is called a "soft" hand. Otherwise, it is called a "hard" hand. You compute the score of a soft hand by counting the ace as 11. For example, the hand five, three, ace is "soft 19." The hand seven, nine, ace, is "hard 17" because if you count the ace as 11, the score exceeds 21. You can't bust by drawing to a soft hand. This doesn't mean you should always draw to a soft hand. For example, the hand four, five, ace is "soft 20," a good hand. In many casinos, the dealer must draw with soft 17.

Doubling Down.
After being dealt the first two cards, you have the option of "doubling down," that is, doubling your bet and receiving exactly one more card. To do this, you turn your cards face up and place an equal amount of chips next to your original bet. The dealer then deals you a card face down. Your hand consists of these three cards and will be compared with the dealer's hand after the dealer plays. The bet is twice your original bet.

Doubling down is smart if there's a good chance you will beat the dealer with one more card. For example, if your hand is seven, four, and the deck is rich in 10-value cards, there's a good chance you'll be dealt a 10-value card, giving you 21.

    In Las Vegas, you can double down with any two cards. In most Reno and Lake Tahoe casinos, you can double down only with a score of 10 or 11.

Splitting Pairs.

If your first two cards are the same denomination, you can turn your hand into two hands by "splitting pairs." For example, suppose you are dealt a pair of eights, giving you a score of 16. To split the pair, turn the eights face up, putting one in front of your bet and the other next to it. Put chips equal to your original bet next to the second eight, and tell the dealer you are splitting the pair. The dealer will treat your pair of eights as the beginning of separate hands, dealing to them one at a time, face up. If you split a pair of aces, you only get one card dealt to each ace.

Blackjack.

If your or the dealer's first two cards are an ace and a 10-value card (ten, jack, queen, king), the hand is called a "natural" or "blackjack." A natural is a two-card, soft 21. If you have a natural and the dealer doesn't, you win. The payoff odds for a natural are 1½ to 1.

Insurance.

After the first two cards are dealt, if the dealer's up card is an ace, you can make an "insurance" bet that the dealer has a natural by placing up to half your original bet in the **INSURANCE** area on the table (Figure 12). The dealer asks the players if they want to make this bet. Payoff odds for an insurance bet are 2 to 1. This bet is supposed to give the player "insurance" against losing to a dealer's natural.

    After insurance bets are made, the dealer peeks at her hole card. If she has a natural, insurance bets win, but the dealer's natural beats everyone who doesn't have a natural. Thus if you make an insurance bet and don't have a natural, and the dealer has a natural, you win your insurance bet and lose your original bet. If the dealer doesn't have a natural, you lose your insurance bet and play out the hand. If the dealer and you both have naturals, you win your insurance bet and tie your original bet. And so it goes.

    As with any bet, you shouldn't make an insurance bet unless you have positive expected winnings. You win an insurance bet when the dealer's hole card is a 10-value card. With a full deck, 16 cards are 10-value cards. If the dealer's up card is an ace, 16 of the remaining 51 cards are 10-value cards, so (disregarding other cards you have seen) the chance that the dealer's hole card is a 10-value card equals 16/51, or .31. The payoff odds on an insurance bet are 2 to 1, so your chance of winning has to be greater than $1/3 = .33$, for the bet to be favorable (for fair odds 2 to 1, $P(\text{win}) = 1/3$). Don't make an insurance bet when the deck is full, since the fraction of 10-value cards equals .31, less than .33. In general, don't make an insurance bet unless you know that the fraction of 10-value cards remaining in the deck is greater than 1/3.

Surrender. Some casinos allow you to "surrender" your hand, quitting after the first two cards are dealt. When you do this, you lose half your original bet.

Multiple Decks. Some blackjack games are played with a single deck, some with two, some with four, some with five, and some with six decks. Four-deck games and higher are dealt from a "shoe," a plastic box with an opening in the bottom from which the dealer slides the cards.

In single-deck games, after each hand the dealer puts used cards face up on the bottom of the deck. At some point, usually near the end of the deck, the dealer shuffles the cards. The dealer can shuffle any time. In multiple-deck games, a marker is placed in the deck. When the marker is reached, the cards are shuffled. Some high-tech dealing shoes have automatic shuffling mechanisms that shuffle cards as they are replaced in the shoe.

Single-deck games, multiple-deck games, what's the difference? Multiple-deck games were introduced to make it more difficult for black-jack pros to keep track of the cards as they were dealt. There's a positive feature to games dealt from a shoe. It's difficult for the dealer to cheat.

Minimum and Maximum Bets. A minimum bet sign is posted at every table. Different tables have different limits. Typically, the minimum bet is from $2 to $5 and the maximum from $500 to $2000. In most casinos there are a few tables reserved for big bettors, with minimum bets ranging from $25 to $100.

A BASIC STRATEGY FOR
WINNING AT BLACKJACK

Computer simulations in the 1950s produced blackjack strategies that provide an even game or give the player a slight edge. All winning strategies are based on the rule that requires the dealer to draw with a score of 16 or less, while the player can draw or stand at any time. The player's advantage comes from situations where the dealer is "forced" to bust. For example, if the dealer has a three, four, five, or six showing, she is likely to bust when drawing if she has a high-value card in the hole. In this situation, the player should be conservative about drawing so as not to bust first. If the dealer has a high-value card showing, she is likely to have a good hand, and the player should be bolder about drawing.

In addition, the chances of winning are increased by keeping track of the cards as they are dealt. "Card counting," as this is called, requires an additional level of skill. When the deck is rich in 10-value cards, the hand is advantageous to the player, who should make relatively large bets. When the deck is rich in low-value cards, there is less chance that the dealer will bust, and the player should make smaller bets. With a card counting system, the player can keep track of the distribution of cards in the deck.

It takes concentration and dedication to perform the tedious task of card counting. Take my advice: Be happy with the basic strategy described here. If you insist on learning card counting, read Edward O. Thorp's classic *Beat the Dealer*, or *The World's Greatest Blackjack Book* by Lance Humble.

Standing and Drawing. The first part of a basic strategy (no card counting) is for standing and drawing with a hard hand.

You should always draw, double down, or split pairs if you have a hard hand with score 11 or less.

If you have a hard hand with score 12 or more, there are three rules to remember:

1 – Draw if your score is 12 through 16 and the dealer's up card is seven or higher.

2 – Stand if your score is 17 or more.

3 – Stand if your score is 12 through 16 and the dealer's up card is two through six.

Next, standing and drawing with soft hands. There are two easy rules to remember:

1 – Stand if you have soft 19 or more, or if your score is soft 18 and the dealer's up card is eight or less.

2 – Draw if you have soft 17 or less, or your score is soft 18 and the dealer's up card is nine, ten, or ace.

Doubling Down. Doubling down takes precedence over drawing. For doubling down with a hard hand (neither of your first two cards is an ace), there are three ever-so-easy rules:

1 – Double down if your score is 11 and the dealer's up card is 10 or less.

2 – Double down if your score is 10 and the dealer's up card is 9 or less.

3 – Double down if your score is 9 and the dealer's up
card is four, five, or six.

If none of these rules applies, follow the previous instructions for
hard drawing and standing.

For doubling down with a soft hand (one of your first two cards is an
ace), follow these rules:

1 – Double down if you have "ace, two" or "ace, three,"
and the dealer's up card is five or six.

2 – Double down if you have "ace, four" or "ace, five,"
and the dealer's up card is four, five, or six.

3 – Double down if you have "ace, six" or "ace, seven,"
and the dealer's up card is three through six.

Splitting Pairs. Here are the effortless rules for splitting pairs:

1 – Always split a pair of aces or eights.

2 – Never split a pair of fours, fives, or tens.

3 – Split a pair of twos or threes when the dealer's up
card is four through seven.

4 – Split a pair of sixes when the dealer's up card is three
through six.

5 – Split a pair of sevens when the dealer's up card is two
through seven.

6 – Split a pair of nines when the dealer's up card is two
through six, eight, or nine.

If none of these conditions are met, follow the rules for doubling
down, standing, and drawing.

Insurance. As I mentioned earlier, the insurance bet gives you positive
expected winnings if the fraction of 10-value cards left in the deck(s) is
greater than 1/3. If you aren't counting cards, don't make an insurance
bet.

Surrender. If your first two cards are nine, seven, or ten, six, surrender
if the dealer's up card is ace, ten, or nine. If your first two cards are nine,
six or ten, five, surrender if the dealer's up card is ten.

How Much Should You Bet?

First, determine your betting unit (minimum bet). The amount of your betting unit depends on your gambling bankroll, that is, how much you have to bet with. The Kelly system (see Chapter 5) can be used to determine your betting unit for a given bankroll. If you can use the basic strategy perfectly, it is reasonable to make your betting unit 1% or 2% of your bankroll. For example, if you have a bankroll of $500, make $5 or $10 bets. If you have $5000 to gamble with, make $50 unit bets. If you win big, increase your betting unit. If you lose big, go sit in a Jacuzzi.

CARD COUNTERS VERSUS THE CASINO

When Thorp's book came out, panicked casinos tried to counter the counters in various ways, such as ejecting them from the casinos. With the aid of one-way mirrors above the tables (known as the "eye in the sky"), photos were taken to identify system players. Card counting isn't illegal, but who wants to argue legalities with a burly goon who politely (for the time being) asks you to leave the premises?

Card counters responded to the casino crackdown in various ways. One bettor I know who had been thrown out of every major casino in Nevada decided to completely change his appearance. Dave, a pasty-faced gambler who went outdoors only to go from one casino to another, grew a beard, dyed his hair, went to a tanning salon, and bought a flashy suit. He tested his disguise on his friends. His brother didn't recognize him. He went to a Las Vegas casino and sat down at a blackjack table. Five minutes later, the pit boss approached him and said, "Hi, Dave. Could I interest you in taking a walk out the front door?" When Dave asked the pit boss how he recognized him, the pit boss just pointed to Dave's watch. Dave had forgotten to remove the expensive watch his girlfriend had given him a few years earlier.

One method used by casinos to detect card counters is to watch betting patterns. An average bettor won't make a sequence of $5 bets followed by a few $100 bets, then back to $5 bets. One gambler I know was ejected from a Lake Tahoe casino merely for varying his bets. Ironically, he was $1000 down when he was nabbed.

Another professional gambler, Lou, invented an elaborate strategy for varying his bets without getting caught. He called it the "Big Player" concept. Lou organized three teams of blackjack players, with five players on a team. Each team consisted of four card counters and a Big Player. The team would enter a casino one by one, the card counters sitting at separate blackjack tables. The counters counted cards and made minimum bets. Meanwhile, the Big Player swaggered through the casino like a drunken high roller. When the count at a table got high, the card counter at this table would signal the Big Player, who would lurch over and make big bets. If the count got low, the counter would signal again, and the Big Player would leave the table. The counters were always making small bets, and the Big Player was always making big bets. No bet variations! It took a year for the casinos to catch on. When they did, Lou's blackjack teams had to leave town in a hurry.

The Card Counting Computer.

When Lou's Big Player teams were disbanded, he hit on another scheme for beating the casinos. He hired a computer expert to develop a card-counting computer small enough to be taken into a casino. The computer was strapped around the bettor's thigh. Small sensing devices were placed under the bettor's toes for tapping information into the computer as cards were dealt, and a buzzer signaled the bettor to draw, stand, etc., as well as how many units to bet.

Lou's computer was not a success. Although it worked fine in the lab, people had trouble using it in a casino environment. Lou later tried hiding a similar device in a pair of very large sunglasses. A pit boss soon discovered it and booted out the user.

THE LOLLAPALOOZA

As Lou will attest, even expert blackjack players can have trouble making a buck. The same can be said for other presumed games of skill, as the following cautionary tale shows.

A professional gambler vacationing in Montana stopped at a local tavern to have a beer. A poker game was in progress, so the gambler decided to take the local cowboys for a few dollars. After playing for a while, the gambler was dealt a flush. The betting was heavy. Soon there was $5000 in the pot and one cowboy left betting against the gambler. When the players showed their hands, the cowboy had 2, 4, 6, 8, 10, two hearts, three clubs. The gambler laughed, showed his flush, and started collecting the pot. The cowboy stopped him. "You lose," said the cowboy. "I've got 2, 4, 6, 8, 10, two hearts, three clubs. That's a Lollapalooza. In this town, a Lollapalooza beats anything."

Looking around the room, the gambler decided not to argue with the cowboys. He figured he'd recoup his losses, which he did in an hour. Then, the gambler was dealt 2, 4, 6, 8, 10, two hearts, three clubs—a Lollapalooza. The gambler bet big. Soon, there was $20,000 in the pot and one cowboy left in the hand. The cowboy showed a full house. "Sorry," said the gambler, as he started raking in the pot. "This time, I've got the Lollapalooza."

"Not so fast, podner," said the cowboy as he grabbed the gambler's arm. "Only one Lollapalooza per night!"

STATE LOTTERIES

Suckers have no business with money anyway.

—CANADA BILL JONES, LEGENDARY
THREE-CARD MONTE DEALER

Lotteries are not a new idea. The Bible describes casting lots to divide up land. Augustus Caesar had a public lottery to raise funds for the restoration of Rome. In the sixteenth century, Italian public lotteries paid cash prizes. The Irish Sweepstakes has been played in this country for years. The Northwest German State Lottery, started in 1612, is now in its 377th year and has the feature that "one of every three tickets is a winner."

In America, state lotteries are becoming an increasingly popular way to raise revenue. In this chapter I'll discuss the California Lottery, where "the schools win, too."

The California Lottery has two main games, Scratch-off and Lotto. Two other games have been recently added, Decco and Topper. I'll discuss these games in detail. The California Lottery is similar to other state lotteries. To put it another way, if you've seen one state lottery, you've seen them all.

SCRATCH-OFF

You buy a ticket for $1 and scratch off an opaque coating to see if you win. Prizes vary from a $1 (money back) return to a chance to participate in the "Big Spin," where you spin a wheel of fortune and win up to $2 million. For promotional reasons, the game changes every six weeks or so, but the payoff structure remains essentially the same. Each game has a catchy name and accompanying hype. Here are some examples of scratch-off games that have been held in the past.

In Lucky Streak, you "win up to $10,000 or a chance at the Big Spin instantly." In 7-11-21, "It can all add up to $21,000." In Wild Card, "the Joker's gone wild with more ways to win in California's newest instant game." In Eureka, you "discover the easiest way to the Big Spin." In Cash Register, you "ring up some instant winners." In Win and Spin, "One scratch could send you spinning." There is extensive advertising to attract bettors. Robin Leach (*Lifestyles of the Rich and Famous*) has appeared on TV spots to pitch these games (*Lifestyles of the Poor and Desperate?*).

In instant scratch-off games, the state takes about 50% of the action, 34% going to public education and the rest covering administrative and promotional costs. The remaining 50% is returned in prizes. Since the state keeps 50 cents of every dollar bet, the house edge is 50%, making the scratch-off game worse than any game in Nevada. Why do people play this game? It's the only game in town.

Win and Spin. Win and Spin was a typical California Lottery instant scratch-off game. In this game, if a ticket had three matching prize amounts, you won that prize. If it had three matching "spins," you went to the Big Spin, a weekly televised event.

There were 135 million tickets printed for Win and Spin, with 22,768,623 winners. This made the chance of winning a prize equal 22,768,623/135,000,000, or about 1 in 6. Almost 11 million of the instant prizes, or 47%, were $1 prizes. Since a ticket cost $1, if you got a $1 prize, you got your money back. Here are the Win and Spin payoffs and the number of tickets of each type.

Instant Prize	No. of Winners
$ 1	10,800,000
2	8,100,000
5	3,240,000
10	540,000
50	54,000
100	27,000
500	6,073
1,000	1,350
10,000	150
Big Spin	50

To find the probability of winning a particular prize, take the number of tickets for that amount and divide by the total number of tickets

(135 million). For example, the chance of winning a $1 prize is 10,800,000/135,000,000, or about 1 in 12. The chance of winning a $50 prize is 54,000/135,000,000, or about 1 in 2,500. The chance of winning a $500 prize is 6,073/135,000,000, or about 1 in 22,000. The chance of winning a $1000 prize is 1 in 100,000. The chance of winning a $10,000 prize is 1 in 900,000.

You go directly to the Big Spin by getting one of the 50 Big Spin winners (the chance is 1 in 2.7 million). If you win a $500 prize, you are entered in a drawing, which also gives you a chance to go to the Big Spin. Your overall chance of going to the Big Spin is about 1 in 2 million.

To find your expected payoff from instant prizes, not counting the Big Spin, multiply each payoff by the number of winners, add, and divide the resulting sum by 135,000,000. Equivalently, multiply the prize amounts by their probabilities (number of winners/135,000,000) and add. The result is an expected payoff of about 44 cents. Subtracting the dollar you paid for the ticket leaves you with expected winnings = −$.56.

The Big Spin. You spin the wheel of fortune and win a prize ranging from $10,000 to $1 million, depending on where the pointer comes to rest. There are 100 sections on the wheel. The Big Spin payoffs are as follows.

. . . .

Prize	No. of Sections
$ 10,000	10
20,000	16
30,000	16
40,000	16
50,000	16
100,000	10
1,000,000	8
Double	8

. . . .

If the pointer ends up on one of the "double" sections, you spin again, and your prize is doubled. (If you hit a "double" section more than once, you keep spinning, but your winnings are still only doubled.) Expected Big Spin winnings, including double, are about $133,000. Since the chance of making it to the Big Spin is 1 in 2 million, your expected Big Spin payoff at the time you buy a ticket is $133,000 divided by 2 million, or about 6 cents.

Adding this to the 44-cent expected payoff from instant prizes brings the overall expected payoff from a single ticket to 50 cents, making the house edge 50%. Ouch!

Here's another way to look at it. The entire prize payout is about $68 million. If you bought every ticket for this game, you would win it all. You would win every instant prize and be the only person on the Big Spin. It would cost you $135 million to buy every ticket. Thus, you would end up losing $135 − $68 million = $67 million, about 50% of your investment. Of course, you would also be contributing to education (and the billion dollar industry that prints tickets and runs the game).

LOTTO

Lotto is similar to keno. You buy a $1 ticket and pick six numbers from 1 to 49. Your ticket is good for one drawing, in which six winning numbers and a "bonus" number are randomly selected from the numbers 1 to 49.

If 3 of your numbers are among the 6 winning numbers, you win $5. If 4 of your numbers are among the 6 winning numbers, you split the prize pool with everyone who also has 4 matches. If 5 of your 6 numbers are among the 6 winning numbers and your sixth number isn't the bonus number, you split the prize pool with the other winners in that category. If 5 of your 6 numbers are among the 6 winning numbers and your sixth number is the bonus number, you split the prize pool with the others in that category. Matching 6 for 6 is called the "jackpot." If you win the jackpot, you split it with any others who picked the same 6 numbers.

Splitting prize money with the winners means that there are no fixed payoff odds. This payoff system, in which payoffs are determined by how much is bet *and* how many winners there are, is called a "pari-mutuel" payoff system.

If no one wins the jackpot (or other prize), the money is added to the jackpot for the next drawing. Thus, there are occasional large jackpots.

As with instant scratch-off games, 50% of the money bet on Lotto is kept by the state and 50% is returned in prizes. In a sense, the house edge is 50%. However, since winners split the pot, and since the jackpot grows when no one wins, expected winnings can't be computed exactly. The chances of winning the various prizes, along with the percentage of the prize pool allocated to each prize, are as follows.

CALIFORNIA 6 FOR 49 LOTTO ODDS AND PRIZES

Match	Chance of Winning	% of Prize Pool
All 6 (jackpot)	1 in 13,983,816	40
5 plus bonus	1 in 2,330,636	21.35
Any 5	1 in 55,491	11
Any 4	1 in 1,032	10
Any 3 ($5 payoff)	1 in 56	17.65

Chances of Winning. To find the chances of winning Lotto prizes we use the combinations formula given in Chapter 4. The probabilities are similar to keno probabilities and are computed as follows.

To match 3 of 6, 3 of the 6 numbers you pick must be among the 6 winning numbers, and the other 3 must be among the 43 nonwinning numbers. There are $\binom{49}{6}$ ways to choose 6 numbers from 49. There are $\binom{6}{3}$ ways to choose 3 numbers from 6, and $\binom{43}{3}$ ways to choose 3 numbers from 43, so the chance of matching 3 of 6 is

$$\frac{\binom{6}{3}\binom{43}{3}}{\binom{49}{6}} = .0176504 = 1 \text{ in } 56$$

To match 4 of 6, 4 of your 6 numbers must be among the 6 winning numbers, and the other two must be among the 43 nonwinning numbers. There are $\binom{6}{4}$ ways to choose 4 numbers from 6 and $\binom{43}{2}$ ways to choose 2 numbers from 43, so the chance of matching 4 of 6 is

$$\frac{\binom{6}{4}\binom{43}{2}}{\binom{49}{6}} = .0009686 = 1 \text{ in } 1032$$

To match 5 of 6 but not the bonus number, 5 of the 6 numbers you pick must be among the 6 winning numbers, and the sixth must be among the 42 that are neither among the 6 winning numbers nor the bonus number. There are $\binom{6}{5}$ ways to choose 5 numbers from 6 and $\binom{42}{1}$ ways to choose 1 number from 42, so the chance of matching 5 of 6 but not the bonus number is

Play Lotto 6/49. Pick the winning numbers and you could win millions. How many millions? That all depends on how many people play each week. And how many pick the right numbers.

The playing is easy.

Pick your six lucky numbers. They could be the ages of your children. Or their birth dates. Even the numbers on your driver's license. Any six different numbers will do as long as they are between 1 and 49.

Now, take a Play Slip. Your Lotto 6/49 retailer will have plenty of Play Slips. On each Play Slip there are five Game Boards. With a heavy vertical mark, fill in the numbered boxes that correspond with your "six lucky numbers" on a Game Board.

Next, give the Play Slip, along with a dollar for every Game Board you have marked, to the retailer. In return he'll give you a ticket that lets you know your selection has

been entered in the weekly Lotto 6/49 drawing. Check your ticket to make sure it has the six numbers you selected. Sign the back of your ticket. After you have signed it no one else can cash it in. Signing your ticket is the best way to protect it.

You can play Lotto 6/49 wherever you see a Lotto 6/49 sign. And you can play

anytime the store is open between 6 a.m. and 10 p.m.—right up to five minutes before the weekly Saturday night drawing.

And the winner is….

When you play Lotto 6/49 you won't have to wait very long to find out if you've won. Every Saturday night there will be a live televised drawing of the six winning numbers. Check your local listings for the time and station.

If you pick the right six numbers, in any order, you win the Jackpot. Probably millions.

You also can win by picking any five of the six. Or four. Even matching three of six will make you a winner.

If you miss the televised Lotto 6/49 drawing, you'll find the winning numbers posted wherever Lotto 6/49 tickets are sold.

FIGURE 14

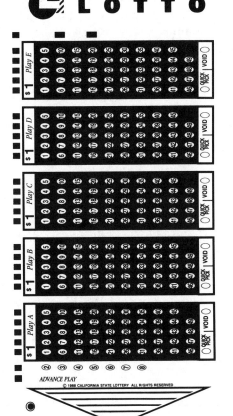

PLAY LOTTO

- **Select 6 numbers** from 1 to 49 for each play.
  - ▸ Place a vertical mark in each chosen number spot.
  - ▸ **DO NOT USE RED INK.**
- Or mark the Quick Pick spot and the computer will pick your numbers.
- Play up to five times on each playslip.
- Play additional draws by marking Advance Play.
- If you make a mistake, mark the void spot. **DO NOT ERASE.**
- Pay the retailer $1.00 for each set of six numbers marked and hand in your playslip.
- You will receive a ticket which lists your chosen numbers (or Quick Pick numbers) and draw date(s).
- Check your ticket immediately to ensure that the numbers are those you've chosen.
  - ▸ Remember the ticket is the only valid receipt.
  - ▸ Sign the back for safety.
- Watch for the winning numbers on TV on each draw date of your ticket.
- For winning number information call:

ENGLISH	SPANISH
976-4CSL	976-5CSL

25¢ OR LESS PER CALL PLUS LONG DISTANCE CHARGES IF REQUIRED.

HOW TO CLAIM YOUR PRIZE

- Present a winning ticket to any CSL **Lotto** retailer for validation.
- A validated winning ticket of up to $99.00 will be paid by the retailer. These prizes may be claimed at the retailer within 180 days after the draw date on the ticket.
- A winning ticket larger than $99.00 must be validated by a retailer. The retailer will present claimant with a claim form, a claim ticket, and the original ticket. The winner submits these three items to CSL for payment. The form and receipt must be postmarked or received by CSL within 180 days after the draw date on the ticket.

PRIZE DIVISIONS AND ODDS OF WINNING

MATCHING NUMBERS	ODDS
6 of 6 Numbers*	1 in 13,983,816
5 of 6 plus Bonus Number	1 in 2,330,636
5 of 6 Numbers	1 in 55,491
4 of 6 Numbers	1 in 1,032
3 of 6 Numbers**	1 in 56

*Lotto winners of $700,000 or more may be paid in equal annual installments for up to a twenty (20)-year period.
**Pays $5.00; all other payouts are calculated by a pari-mutuel formula.

- Tickets, transactions and prize payments are subject to State Law and the rules and regulations of the California State Lottery.
- Upon submission of a Lotto playslip for processing, the player agrees to abide by the California Lottery rules and regulations.
- This playslip is not a valid receipt.

LOTTO is governed by state law and the rules and regulations of the California State Lottery. State law prohibits the sale of a lottery ticket or the payment of a prize to a person under the age of 18 years.　　　LO-5/88 1B

FIGURE 15

$$\frac{\binom{6}{5}\binom{42}{1}}{\binom{49}{6}} = .0000180 = 1 \text{ in } 55{,}491$$

To match 5 of 6 plus the bonus number, 5 of the 6 numbers you pick must be among the 6 winning numbers, and the sixth must be the bonus number. There are $\binom{6}{5}$ ways to choose 5 numbers from 6 and only 1 way the remaining number can be the bonus number, so the chance of matching 5 of 6 plus the bonus number is

$$\frac{\binom{6}{5}}{\binom{49}{6}} = .0000004 = 1 \text{ in } 2{,}330{,}636$$

There are $\binom{49}{6} = 13{,}983{,}816$ ways to choose 6 numbers from 49, and only one way you can match 6 of 6, so the chance of winning the jackpot is 1 in 13,983,816.

Someone who heard me discussing Lotto odds on a talk show called in and said that the Lottery Commission had made a mistake in computing odds for the 5 of 6 (without bonus number) prize. He said that he had personally called the Lottery director and tried to bet him $1000 that the odds stated on the Lotto brochure were wrong. After a lengthy discussion, I convinced the caller that the stated odds were correct. There was a long silence. Finally, he said in a whisper, "Then why didn't the Lotto director take my bet?"

Payoffs. Suppose $20 million is bet on a Lotto game. Then 50%, or $10 million, is kept by the state, leaving $10 million in prize money. The payoff pool is allocated as follows:

Category	% of Total	Amount
Any 6 (jackpot)	40	$4,000,000
5 plus bonus	21.35	2,135,000
Any 5	11	1,100,000
Any 4	10	1,000,000
Any 3 ($5 payoff)	17.65	1,765,000

The chance of winning the jackpot is about 1 in 14 million, so there are typically 1 or 2 jackpot winners when $20 million is bet (20 million tickets purchased). The chance of matching 5 plus the bonus equals 1 in 2,330,636, so there are about 8 or 9 such winners when $20 million is bet (20 million divided by 2,330,636), making a typical 5 plus bonus payoff about $2,135,000 (5 plus bonus prize pool) divided by 9, or $237,222. Since the chance of matching 5 of 6 without bonus is 1 in 55,491, there are about 360 such winners for every 20 million tickets, making a typical 5 of 6 prize payoff about $1,100,000/360, or $3,055. The chance of matching 4 of 6 is 1 in 1032, so there are about 19,380 such winners, making a typical 4 of 6 prize payoff equal $1,000,000/19,380, or $52. The chance of matching 3 of 6 is 1 in 56, so there are about 357,143 $5 winners ($1,765,000/357,143 = $4.94).

What the Jackpot Odds Really Mean.

The chance of winning the jackpot is about 1 in 14 million. To put these odds in perspective, if you buy 50 Lotto tickets a week, you'll win the jackpot about once every 5000 years. If you have to drive 10 miles to buy a Lotto ticket, you're three times more likely to get killed in a car crash on your way to buy the ticket than you are to win the jackpot. If your car gets 25 miles per gallon, and you purchased a gallon of gas for every Lotto ticket you bought, you would buy enough gas for approximately 730 round trips to the moon before you won the jackpot. Suppose you're in a football stadium filled with 70,000 people, and there are 200 such stadiums. Select one person at random from these 200 stadiums. Your chance of being selected is the same as your chance of winning the Lotto jackpot. As one gambler put it, "Your chances of winning the jackpot are about the same, whether or not you buy a ticket." So what? By the law of very large numbers, if enough people buy tickets, someone will win.

Occasionally, there are a series of games with no jackpot winners, causing the pot to get very large. When the jackpot gets large, more people bet. The biggest jackpot to date was $67,028,495, on February 21, 1990. Four winners split that amount. The largest single payoff was $25,140,000, on December 16, 1987.

The Annuity System.

Big jackpot payoffs—prizes of $1 million or more—are paid over 20 years and thus cost the Lottery Commission about half of the face value of the prize. For example, if you win a $10 million jackpot, the commission needs to invest only about $5 million— what is known as the "present value" of the jackpot—to generate annual payments totaling $10 million over 20 years.

Lotto Systems.

The California Lottery is a billion-dollar business. There are numerous "experts" who sell "winning" Lotto systems. Publications advertise systems for "beginners," "advanced" systems, and systems based on astrology, numerology, and various other dubious principles.

A Los Angeles newsletter keeps track of Lotto numbers and tries to find predictable patterns. The newsletter offers a "computer analysis" and a choice of strategies. Right next to a list of the "Hot 10 numbers," numbers that have recently "hit 5 or more times," is a list of the "25 Most Probable," those numbers that "after hours of analysis . . . may be due." If you pick from the Hot 10 list, you'll be choosing numbers that have come up often in the past. If you select one of the 25 Most Probable, you're opting for numbers that *haven't* come up very often. Is there a contradiction? Maybe you should pick some of each.

The real truth about lotto systems is that the Lottery Commission monitors the randomness of the selection methods, and there's no way to predict randomly selected numbers. Put simply, trying to predict Lotto numbers is a waste of time.

Is There Anything You Can Do?
Although there's no way to improve your chances of winning, there *is* something you can do to make your expected winnings larger than if you pick numbers at random. Pick numbers unpopular with the betting public. If you choose numbers that no one else picks, your chances of winning are still the same, but if you win, you don't have to split the pot with anybody. The trick is to determine which numbers are unpopular.

In a study of Canadian Lotto (same setup as California), statisticians Tom Cover and Hal Stern demonstrated that picking unpopular numbers can improve your expected jackpot winnings. Using a mathematical model and data available for the Canadian Lotto, Cover and Stern showed that the best ticket for maximizing expected payoff (by minimizing the number of people splitting the jackpot) was the "least popular" ticket, consisting of the numbers

20, 30, 39, 40, 41, 48

The "most popular," or worst, ticket consisted of the numbers

3, 7, 9, 11, 25, 27

For a typical jackpot, selecting the least popular ticket increased a player's expected payoff by more than 50% over a randomly selected ticket. Alas, this wasn't enough to provide a favorable bet. Expected payoff was still less than the cost of the ticket!

The least popular ticket is based on Canadian Lotto data. There is no assurance that Californians bet the same way. If enough people read this book and bet on the least popular ticket, it won't be the least popular ticket any more. Therefore, I'll tell you the *second* least popular ticket:

20, 30, 39, 40, 48, 49

One thing you can do to *guarantee* winning the jackpot is to buy all possible ticket combinations. Since there are 13,983,816 ticket combinations, it will cost you $13,983,816. If the jackpot is greater than that, you will be certain of a profit — unless, of course, you have to split the jackpot with other winners. In that case, you will be in deep trouble. There's also a logistics problem. If you filled out three tickets a minute and worked 24 hours a day, seven days a week, it would take about nine years to write all the ticket combinations.

The American Dream. I was on a talk show explaining Lotto odds when an outraged caller informed me that I had no business criticizing the lottery. She said that the possibility of going from rags to riches was part of the American Dream and that Lotto was a way to make this dream come true. I agreed with her that going from rags to riches is part of the American Dream, but *another* part of the Dream is that if you are clever and work hard, you can increase your chance of getting the riches. Lotto involves no skill.

On the other hand, Lotto provides a nice mechanism for randomly distributing wealth. There are an average of 200 new millionaires a month created by state lotteries.

The New, Improved California Lotto. When Lotto was first introduced, bettors recognized its redeeming features. It was cheap, legal, and the only game in town. After a series of public relations blitzes, ticket sales leveled off at a few million per game. Since it takes an average of 14 million tickets to produce a jackpot winner, there were many games with no jackpot winner. (Luck is a group activity.) Jackpots grew. With big jackpots, more tickets were purchased. Every time there was a big jackpot and ticket sales went up, sales stayed up in future games.

After a number of big jackpots, ticket sales climbed to about 20 million per game. Enter the law of very large numbers. Since the chance of winning the 6 in 49 Lotto jackpot is 1 in 14 million, with 20 million tickets there's likely to be a winner. When there's usually a winner, jackpots don't get big and ticket sales don't increase. Enter the Lottery Commission. In the spring of 1990, 4 more numbers were added to the California Lotto. Now you choose 6 numbers from 53 instead of 6 from 49. The commission did this to make the odds worse (23 million to 1) and create more big jackpots, causing more ticket sales.

A problem arises. If tickets sales go up as planned for the new Lotto game, there will soon be enough players to have frequent jackpot winners again. Sales will level off, and, except for the new crop of suckers, things will be back to where they started. The conclusion is inescapable: The game will have to be changed again! And so on.

For example, suppose that because of future increased ticket sales, the Lotto commission changes the game from 6 in 53 to 14 in 50. (Instead of picking 6 numbers from 53, you pick 14 from 50.) Then the chance of winning the jackpot would be about 1 in a trillion. If 50 million tickets were sold for each 14 in 50 Lotto game, with two games per week there would be a jackpot winner on the average of once every 192 years. Maybe it's worth the wait. The jackpot would be huge.

This is a challenge for Lotto promoters. If each of the 5 billion men, women, and children on earth bought 20 tickets per game for 14 in 50 Lotto, there would be a jackpot winner about once every 10 games, guaranteeing big jackpots and more ticket sales. Nothing lasts forever. If everyone on earth bought 300 tickets per game, there would be jackpot winners for most games, and ticket sales would level off.

Eventually, there won't be enough people on the planet to support Lotto growth. I say let's intensify our efforts to explore outer space. There must be intelligent life somewhere in the universe. At least life intelligent enough to buy Lotto tickets.

· · ·

In addition to changing Lotto, the commission added two new games: Decco and Topper.

Topper.

Topper can be played only when you buy a regular Lotto ticket. For an additional dollar, you pick three cities from the hundred largest cities in California. At the Lotto drawing, three winning cities are randomly selected. The payoffs are fixed, not pari-mutuel. If your three cities are the same as the three winning cities, you get $25,000. If two of your three cities are selected, you get $95. If one of your three cities is selected, you get $2. You obtain the various win probabilities with the selections formulas as we did with Lotto probabilities. The win probabilities, expected payoff, and house edge for Topper are as follows:

· ·

Result	Payoff	Probability
3 for 3	$25,000	.0000062 (1 in 161,700)
2 for 3	95	.0018 (1 in 556)
1 for 3	2	.0864 (1 in 12)

· ·

$$E = (25{,}000 \times .0000062) + (95 \times .0018) + (2 \times .0864) = .50$$

Subtracting the $1 cost to play yields expected winnings $= -\$.50$. The house edge $= 50\%$!

Decco.
Decco is separate from Lotto. There are six Decco drawings per week: Monday through Saturday. Sunday, of course, is the day of rest. A religious day. A day to bet on football.

Decco is played with an ordinary deck of 52 playing cards. It costs $1 to play. You pick one denomination from each suit. For example, you could pick ace of clubs, two of hearts, nine of spades, and jack of diamonds. Then, in the televised Decco drawing, one winning card from each suit is selected at random. If your four cards are the same as all four winning cards, you get $5000. If you match three of four, you get $50. If you match two of four, you get $5. If you match one of four, you get a free Decco replay.

You compute Decco probabilities with the multiplication rule for independent events. There are 13 cards for each suit, so the probability that your card in a suit matches the winning card in that suit equals $1/13$. Note that there are $13 \times 13 \times 13 \times 13 = 28{,}561$ four-card combinations, one from each suit. The chance that all four of your cards match the winning card equals

$$(1/13) \times (1/13) \times (1/13) \times (1/13) = 1/28{,}561 = .000035$$

The chance that your card in a suit is not the winning card $= 12/13$. Thus the chance that your cards in three particular suits match the winning cards in those suits and your card in the remaining suit doesn't match the winning card in that suit equals

$$(1/13) \times (1/13) \times (1/13) \times (12/13) = 12/28{,}561$$

Since there are four suits in which your card could be different from the winning card, it follows that

$$P(\text{match } 3) = 4 \times (12/28{,}561) = 48/28{,}561 = .0017 = 1 \text{ in } 595$$

The chance that your cards in two particular suits match the winning cards in those suits and your cards in the other two suits don't match the winning cards in those suits equals

$$(1/13) \times (1/13) \times (12/13) \times (12/13) = 144/28,561$$

Since there are $\binom{4}{2} = 6$ ways to select two winning suits and two losing suits from four suits, it follows that

$$P(\text{match } 2) = 6 \times (144/28,561) = 864/28,561 = .0303 = 1 \text{ in } 33$$

The chance that your card in a particular suit matches the winning card in that suit and your cards in the other three suits don't match the winning card in those suits equals

$$(1/13) \times (12/13) \times (12/13) \times (12/13) = 1,728/28,561$$

Since there are four particular suits in which you could have a winning card, it follows that

$$P(\text{match } 1) = 4 \times (1,728/28,561) = 6,912/28,561 = .2420 = 1 \text{ in } 4$$

The probabilities, expected payoff, and house edge for Decco are as follows.

Result	Payoff	Probability
4 for 4	$5000	.000035 (1 in 28,561)
3 for 4	50	.0017 (1 in 595)
2 for 4	5	.0303 (1 in 33)
1 for 4	E	.2420 (1 in 4)

Since the payoff for matching 1 for 4 is a free Decco replay, its payoff value equals E, the expected payoff for a Decco ticket (what we are computing). Solving the resulting equation, we find

$$E = (5000 \times .000035) + (50 \times .0017) + (5 \times .0303) + (E \times .2420)$$

or

$$E = \frac{(5000 \times .000035) + (50 \times .0017) + (5 \times .0303)}{.7580} = .54$$

Subtracting the $1 cost to play yields expected winnings $= -\$.46$. The house edge $= 46\%$.

THE BOTTOM LINE

State lotteries are the only game in some towns and not the only game in other towns. They are the worst game in all towns.

CHAPTER

8

HORSE RACING

Well, you understand Unser Fritz is betting one hundred thousand dollars against a thousand dollars that Cara Mia will run in the money, and personally I consider this wager a very sound business proposition, indeed, and so does everybody else, for all it amounts to is finding a thousand dollars in the street. There is really nothing that can make Cara Mia run out of the money, the way I look at it, except what happens to her, and what happens is she steps in a hole fifty yards from the finish when she is on top by ten, and breezing, and down she goes all spread out, and of course the other three horses run on past her to the wire, and all this is quite a disaster to many members of the public, including Unser Fritz.

—DAMON RUNYON, *ALL HORSE PLAYERS DIE BROKE*

Americans have been betting on horse races at least since the Revolutionary War. Racing appeals to all types of people, and the crowd at any major racetrack is, to me, as interesting as the sport itself. Where else can you see a chief executive officer standing next to a wino, a preacher next to a politician, an artist next to a stockbroker, a psychiatrist next to a statistician, people of all races and ages screaming for their horses to win?

Betting on the outcome of a horse race is different from betting at a casino, where the odds are set in advance. Payoff odds at the track are determined by how much is bet on each horse. The pari-mutuel system, as this is called, was introduced in the United States after racing scandals involving independent bookmakers led to a shutdown of racetracks in every state but Kentucky and Maryland. Pari-mutuel betting was an immediate success, and it didn't take legislators long to claim a sizable share of track revenues.

With a pari-mutuel payoff system, the "house"—in this case the track operation and state and local governments—take a fixed percentage (the "take") of the betting action and the winning bettors split what's left. The take at U.S. racetracks is typically 17%. That is, for every $100,000 wagered, the house takes $17,000; the remaining $83,000 is available to

113

be divided among winning bettors. Prize monies ("purses"), which are independent of the betting action, are given by the track to the owners of horses that come in first, second, and third (and sometimes fourth) in each race.

Until recently, horse racing was the biggest gambling activity in the country. Football betting is now bigger, but billions of dollars are still wagered annually on horse racing. Tout sheets are published daily. Software is written. Systems are invented. The track welcomes all. They take their 17% off the top. Winning systems (if they exist) take advantage of betting trends, not the track.

The sports sections of most big city newspapers carry write-ups for the day's races at nearby tracks. There is a summary of each race, including names of horses and jockeys, post positions (the order, from the inner rail outward, in which the horses line up to start the race), comments about each horse, weights carried, and probable odds (Figure 16). The probable odds are estimates. Actual payoff odds aren't determined until all bets are placed.

TYPES OF BETS

There are three main bets: win, place, and show. In addition, there are "exotic" bets involving more than one race or more than one horse in a particular race. The minimum bet at most tracks is $2. (Some tracks permit $1 bets on exotics.) Available bets differ slightly from track to track. The following are typical:

Win. When you bet on a horse to win, you win if your horse finishes first.

Place. When you bet on a horse to place, you win if your horse finishes first or second.

Show. When you bet on a horse to show, you win if your horse finishes first, second, or third.

Exacta. To win in exacta wagering, you must pick the two horses finishing first and second in a race, in exact order.

Quinella. You must pick the first two finishers of a race. It doesn't matter which one comes in first and which one comes in second. You win either way.

BAY MEADOWS/Handicap

By Larry Stumes
Chronicle Staff Writer
Saturday Dec. 30 Clear and fast Post time 1 p.m.

837 — FIRST. (1st half DD/Pick9 Begins) six furlongs, three year olds and up, claiming $20,000-18,000, purse $12,000.

PP	Horse	Jockey	Wt	Comments	Odds
7	Island Day Brek	R Hnsen	117	Won three, claimed gain	2-1
2	Macho Cmach	J Lambert	119	Caught similar despite trouble	5-2
4	Calehuche	O A Martinez	117	Stretch-runner back in claimer	3-1
3	G T The Wndw	A Pttersn	117	Beaten neck for $16,000	9-2
5	Zar Moro	R Schacht	117	Lacked win spirit in '89	6-1
6	Doctor Dakota	E Sanchez	117	Improved despite stepping up	8-1
1	Woomera	C Schvaneveldt	117	Trailed richer after rest	12-1

839 — THIRD. ($2 Picksix Begins) abt 1-1/8 mile (turf) fillies three year olds, THE PULLMAN Classified Handicap, purse $14,000.

PP	Horse	Jockey	Wt	Comments	Odds
2	Shady Speclatn	T Chpmn	117	Promising filly before layoff	5-2
7	Chantalong	R Hansen	120	Closed fast to catch these	3-1
1	Pas de Bourree	R Warren	114	Runner-up twice on turf	7-2
6	Caranomi	S Maple	112	Solid return, turf breeding	5-1
5	B' Gotcha	A Gryder	110	Moves up on turf	8-1
3	Ples Of Pleasre	R Gnzlez	116	Shown liking for grass	10-1
4	Mke's Madness	A Castnn	110	Romped over bottom maidens	10-1
8	Rich Cream's Best	J Jdce	112	Once was rich turf filly	12-1

841 — FIFTH. ($2 exacta) six furlongs, three year olds and up, Cal bred, claiming $6250, purse $6000.

PP	Horse	Jockey	Wt	Comments	Odds
11	Fleet Stanley	C Hummel	117	Wired maidens, can repeat	2-1
2	Rocket Rod	K Tohill	117	Cnsistent hrse claimed, rested	3-1
4	Try Boi	E Sanchez	117	Usually runs his race	7-2
1	Agori Luck	D Thomas	117	Forced fast pace in route	5-1
8	Mark The Lrk	B Cmpbell	117	Veteran used to win races	8-1
12	Delaware Ford	A M Ngez	117	Needed last, can improve	10-1
3	Never Dare Me	R Hansen	117	Best races been in mud	12-1
7	Onward Mel	A Gryder	117	Invader 1-for-23 since '87	15-1
9	Wild Pursuit	J Judice	117	Been stopping down south	15-1
10	Uncle Canck	O A Martnez	117	Well-beaten comeback pair	20-1
5	Lcky Edition	R Gonzalez	117	Recent not promising	20-1
6	Big Bad Step	V Miranda	117	Dull trio after win	20-1

Scratched—Last Time Around, Smiley Bo.

000 — SPECIAL. ($2 exacta) six furlongs, all ages, Palos Verdes Handicap (Grade III), purse $107,400. (Santa Anita Simulcast.)

PP	Horse	Jockey	Wt	Comments	Odds
1	Olympic Prospect	A Solis	123	One of nation's best	6-5
6	Sam Who	L Pincay Jr	122	Runner-up at odds-on	8-5
3	Sunny Blossom	G Stevens	115	Speed to battle 'Prospect	5-1
2	Order	R Davis	115	In a bit tough again	8-1
4	Cutter Sam	M Pedroza	112	Toss out turf, 1:39-3 drill	10-1
5	Reconnitering	C McCarrn	112	Doesn't beat this kind	12-1

843 — SEVENTH. ($2 exacta) six furlongs, fillies and mares three year olds and up, THE EASTERN DAWN Overnight Handicap, purse $20,000.

PP	Horse	Jockey	Wt	Comments	Odds
6	Keen Lady	R Schacht	115	Speedy filly worked well	2-1
3	Miss Qlla Illa	T Chapman	116	Rallied to upset similar	5-2
1	Hatti	R Hansen	113	Back to best trip	7-2
2	Abstract Energy	S Maple	114	Needed last, not beaten badly	9-2
5	Lunar Beauty	R Gonzalez	113	Tailed off in routes	5-1
7	Shareen's Pleasre	J B Crz	110	Faced good one in last	8-1
8	Lady James	R Privitera	109	Won 4 of 5 here, faces older	10-1
4	Storm Ryder	I Diaz	112	Beat one in three here	15-1

FIGURE 16

Daily double. To win a daily double bet, you must pick the winners of two designated races, usually the first two races of the day.

Trifecta. To win a trifecta (sometimes called "triple"), you must pick the three horses finishing first, second, and third in a designated race, in exact order.

Pick six. You must pick the winners of six designated races. If no one picks all six winners, payoffs are carried over to the next day less a portion awarded to those who pick the next highest number of winners.

Win Bets.

If you bet on a horse to win, you win the bet if your horse wins the race. If your horse doesn't win the race, you lose. The payoff odds depend on how much is bet on each horse to win.

A race usually has five to twelve horses entered. On a typical day at the track nine or ten races are run, spaced at half hour intervals. A race takes a couple minutes to run. There are about 25 minutes between each race, during which time bettors can try to figure out which horses to bet on, look at the horses, watch the odds, and place their bets. There's also plenty of time for bettors to eat, drink, smoke, and contemplate the ways they could have spent the money they just lost. (I shouldn't be so profit-oriented. People don't always do things just to make money.)

When you place a bet at the betting window, you are given a ticket. The betting windows close about a minute before each race begins.

Let's look at today's seventh race at Upsand Downs. There are five horses entered. The following total amounts are bet to win on each horse:

WIN POOL
. . . .

Big Daddy	$25,000
Prince Charming	$33,000
Lucky Lulu	$18,500
Mr. Ed	$23,000
Morning After	$500

. . . .

The total amount bet on horses to win equals $100,000. This is called the "win pool." The most money is bet on Prince Charming. He is the favorite. Only $500 is bet on Morning After. He is the outsider, that is, the longest longshot.

The amount of the win pool isn't known until all bets are made. A running tally of betting information is shown on the totalizator, or "tote" board, a large computerized display in the track infield. Between races, the tote board is frequently updated, so bettors can see how much has been bet on each horse and the resulting changes in odds.

Okay. The win pool is $100,000. The betting windows are closed. What are the payoff odds to win for each horse? First, the track takes out 17%, or $17,000. This leaves $83,000 for winning bettors to divide among themselves. Suppose Prince Charming wins the race. Since $33,000 was bet on Prince Charming, there is $83,000/33,000 = $2.52 to pay back for each dollar bet. This means if you made a $2.00 bet on Prince Charming, you should get back $2.52 $\times$ 2 = $5.04. Actually, you won't get back quite that much. Payoffs are rounded *down* to the nearest 10 cent increment per dollar payoff. If you bet $2.00 on Prince Charming, instead of being paid $2.52 $\times$ 2 = $5.04, the $2.52 would first be rounded down to $2.50 and you would get $2.50 $\times$ 2 = $5.00. Rounding down to the nearest 10 cents is called "dime breakage" and gives the track (and state) additional revenue, and the winners less profit.

The tote board shows that Prince Charming paid $5.00 to win. If you made a $2.00 bet, you would get back $5.00 — $3.00 in winnings plus the $2.00 you bet. Thus the payoff odds are 3 to 2.

Suppose you bet $2 on Morning After to win, and Morning After wins the race. Since only $500 was bet on Morning After, your payoff per dollar would be $83,000/500 = $166. For a $2 bet, you would be paid $166 $\times$ 2 = $332. This horse is indeed a longshot.

Suppose you bet $40 on Morning After to win, and Morning After finishes second to Prince Charming. You lose.

Suppose you bet $2 on Mr. Ed to win and Mr. Ed wins the race. Since a total of $23,000 was bet on Mr. Ed, your payoff per dollar equals $83,000/23,000 = $3.61, or $3.60 after breakage. For a $2 bet you would be paid $3.60 $\times$ 2 = $7.20. Mr. Ed pays $7.20 to win.

. . . .

Lou, the ingenious blackjack player whose exploits were described in Chapter 6, made a small fortune playing blackjack and lost it all at the racetrack. I asked Lou why he spent so much time playing blackjack only to squander his winnings at the track. With a sad look, Lou said, "Blackjack is a business. Horses are my passion."

Place Bets.

A place bet is a wager that a horse will finish either first or second. The "place pool" in a race is the total amount bet on all horses to place. The place pool is separate from the win pool. Here's the place pool for the seventh race at Upsand Downs.

PLACE POOL

. . . .

Big Daddy	$20,000
Prince Charming	$30,000
Lucky Lulu	$22,500
Mr. Ed	$19,500
Morning After	$1,000

. . . .

The place pool equals $93,000. The most money is bet on Prince Charming, the favorite to win. The tote board displays information on the amount of place bets during the course of betting, as it does for win bets, but shows current odds only for win bets.

The payoffs for each horse to place depend on which two horses place (win or come in second). First, the track takes out $15,810, its 17% cut from the $93,000 place pool. This leaves $77,190 for place winners to divide. Suppose you bet on Prince Charming to place and Prince Charming wins the race. You win your place bet. The amount you are paid depends on which other horse placed, because its backers must also be paid from the place pool.

Suppose Lucky Lulu comes in second. Prince Charming and Lucky Lulu are both place bet winners. The total amount bet on Lucky Lulu and Prince Charming to place is subtracted from the amount left in the place pool after taking out the track cut, yielding $77,190 − $52,500 = $24,690 in profit, which is split equally between place bettors on Prince Charming and Lucky Lulu. A total of $24,690/2 = $12,345 goes to Prince Charming place bettors. The same amount goes to Lucky Lulu place bettors.

Since $30,000 is bet on Prince Charming to place, $2 Prince Charming place bettors get back $12,345/30,000 = .41, which becomes .40 after breakage. Multiplying by 2 and adding the $2 bet gives a payoff of $2.80. The tote board would show that Prince Charming paid $2.80 to place.

Since $22,500 is bet on Lucky Lulu to place, $2 Lucky Lulu place bettors get winnings per dollar bet of $12,345/22,500 = $.55, which becomes $.50 after breakage. Multiplying by 2 and adding the $2 bet gives a payoff of $3.00 for a $2 bet. The tote board shows that Lucky Lulu paid $3.00 to place.

Suppose Morning After, rather than Lucky Lulu, comes in second. Since only $1000 is bet on Morning After to place, there is more losers' money to split than there was with a one, two finish of Prince Charming and Lucky Lulu, and hence the place payoffs are higher. The money to be

split equally between place bettors on Prince Charming and Morning
After is calculated as follows:

$77,190 (place pool after track take) − $31,000 (place bets on
Prince Charming and Morning After) = $46,190

Thus $46,190/2 = $23,095 goes to Prince Charming place bettors and
$23,095 goes to Morning After place bettors.

Since $30,000 is bet on Prince Charming to place, $2 Prince
Charming place bettors get back $23,095/30,000 = $.77, which becomes
.70 after breakage. Multiplying by 2 and adding the amount of the bet
gives ($.70 × 2) + $2 = $3.40. $2 Morning After place bettors get back
$23,095/1,000 = $23.10. Multiplying by 2 and adding the $2 bet gives a
payoff of $48.20.

Show Bets.
A show bet is a bet that a horse will finish either first,
second, or third. The "show pool" in a race is the total amount bet on all
horses to show. The show pool is separate from both the win and place
pools. Here's the show pool for the seventh race at Upsand Downs.

SHOW POOL
. . . .

Big Daddy	$25,000
Prince Charming	$32,500
Lucky Lulu	$35,000
Mr. Ed	$15,500
Morning After	$2,000

. . . .

The show pool equals $110,000. The amount each horse pays to
show depends on which three horses show. The track takes 17% profit
from the show pool, in this case $18,700. This leaves $91,300 for show
bet winners to divide among themselves.

Suppose you bet on Lucky Lulu to show and Lulu finishes third.
Suppose Big Daddy comes in first and Mr. Ed comes in second. A total of
$25,000 + $15,500 + $35,000 = $75,500 is bet on Big Daddy, Mr. Ed,
and Lucky Lulu to show. Since $91,300 remains in the show pool after the
track take, there is $91,300 − $75,500 = $15,800 in losers' money to be
split evenly among the three show winners. This yields $15,800/3 =
$5,266.67 to divide among bettors on each show winner.

Since $35,000 is bet on Lucky Lulu to show, $2 Lucky Lulu show bettors get winnings per dollar of $5,266.67/35,000 = $.15, which becomes $.10 after breakage. Multiplying by 2 and adding the $2 bet yields 2.20. The tote board indicates that Lucky Lulu paid $2.20 to show.

Since $25,000 is bet on Big Daddy to show, $2 Big Daddy show bettors get back ($5,266.67/25,000) = $.21, or $.20 after breakage. Multiplying by 2 and adding the $2 bet gives a $2.40 payoff for Big Daddy show bettors.

Since $15,500 is bet on Mr. Ed to show, $2 Mr. Ed show bettors get back winnings per dollar of $5,266.67/15,500 = $.34, or $.30 after breakage. Multiplying by 2 and adding the amount bet yields a $2.60 payoff.

Exotic Bets. In addition to win, place, and show bets, there are a variety of exotic bets that virtually always have higher payoffs than ordinary bets. The most common exotic bets — daily double, exacta (or perfecta), quinella, trifecta, and pick six — were described earlier. Bets like the pick six wager are not easy to win, and when there are a number of days with no winner, the payoff pool becomes large. This presents an interesting opportunity. For example, suppose the pick six payoff pool has reached $900,000. If there are 8 horses in each of today's pick six races, there are $8^6 = 262,144$ possible ways to pick six winners. It costs $2 \times 262,144 = $524,288 to buy $2 tickets for every possible combination. Since the payoff pool is $900,000, you could bet on every possible combination and guarantee a win. Fine, but if someone else wins, you split the pot and may in fact lose money. If several others win, you could be a big loser.

The track take for some exotic bets is higher than win, place, and show takes. This is unpleasant, but considerably better than the 50% takeout in state lotteries, the other major gambling game with pari-mutuel payoffs.

MINUS POOLS

Suppose there is a race in which a horse is an overwhelming favorite. To use an extreme example, suppose that in the eighth race at Upsand Downs, *everybody* bets on Top Banana to win. If Top Banana wins, there's no losers' money to divide among winners. In fact, if the track were to take its 17%, winning bettors wouldn't even recapture their original bet. This situation is called a "minus pool," and it does occasionally occur. Fortunately, the law requires tracks to pay a minimum amount, even with a minus pool. For example, California and New York tracks are required to pay at least $2.10 for a $2.00 winning bet. Minus pools are the only

way the track (and the state) can get hurt. Don't worry. Be happy. At least
you can't lose money if you win your bet.

TYPES OF RACES

There are four main types of races: maiden races, claiming races, allowance races, and stakes races. There are age, sex, and other qualifications
for entrants in a race.

Maiden races are for horses who have never won a race. Once a
horse wins a race at any recognized track, it can't compete in another
maiden race.

The majority of races are claiming races. Claiming races are struc-
tured to assure that horses entered in a given race are fairly equal in
ability. Every claiming race has a price for which any horse entered in the
race can be purchased up until race time. For example, in a $3500
claiming race, any registered owner can take $3500 to the track office at
any time prior to the race and buy any horse in the race. An owner
typically wouldn't enter a horse worth more than $3500 in a $3500
claiming race because it could be bought for less than it's worth. And an
owner would be unlikely to enter a horse worth less than $3500 in a
$3500 claiming race because it wouldn't have much chance of winning.
Thus the horses entered in a claiming race are all worth about the price of
the race. There are exceptions.

Suppose your horse has been competitive at the $15,000 level but
has an injury no one knows about. You drop it in class and enter it in a
$10,000 claiming race, hoping that some unsuspecting sucker will buy it.
Someone who has seen your horse in its better days thinks it's worth
$15,000 and buys it. The horse hobbles through the race and comes in
last. Someone has just bought a lame horse for $10,000.

You own another horse worth $15,000. It hasn't won recently, but
it's a good horse. You would like to win a race and pick up some quick
purse money. You lower your horse in class and enter it in a $10,000
claiming race, hoping no one will buy it. Nobody buys it. In addition to
purse money, you want to win at the betting windows, so you bet heavily
on your horse, knowing that it can easily outrun $10,000 horses. Another
horse in the race has its legs bandaged. You recognize this horse as having
been in $20,000 claiming races and figure that the owner is trying to
unload it because it's injured. The race begins. The bandaged horse and
your horse quickly outrun the rest of the field. The bandaged horse wins
by a nose. The bandaged horse's owner was trying to pick up easy purse
money, same as you. The bandages were a ruse. This happened to a friend

of mine. As we left the track that dismal day, he said to me, "There must be a better way to go broke." Actually it could have been worse—his horse could have been claimed.

Horses entered in allowance races can't be claimed. The purses for allowance races are larger than the purses for claiming races, and allowance races therefore attract better horses. Each allowance race has conditions that entrants must meet. For example, a typical allowance race might be limited to three-year-olds and older that have never won two races other than a maiden race. "Allowances," in terms of weight to be carried (the jockey's weight plus the "allowance"), are made on the basis of how recently the horse has been successful, as well as its age and sex.

Stakes races are for horses of the highest quality. There are often substantial entrance fees, and added purse money (the "stake") is provided by the track. Weights are added so that all horses of the same age and sex carry the same total weight (including jockey). The Kentucky Derby, the Belmont Stakes, and the Preakness are classic stakes races for three-year-olds.

TRACK INFORMATION

There are four main sources of information for horse players: daily newspapers, the *Daily Racing Form*, the programs sold at the track, and tout sheets.

The best source of data on horse racing is a newspaper called the *Daily Racing Form*, which supplies data on the day's races at every major track in the country. Detailed information is given for each horse in every race, including past performance data, types of races, age, breeder, trainer, lifetime earnings, etc. Figure 17 shows part of the *Racing Form*'s information for the fourth race at Philadelphia Park on October 4, 1990. Serious bettors study the *Racing Form* with the zeal of scholars. They *are* scholars. Some people say it's a mark of distinction to read the *Wall Street Journal* and a mark of degeneracy to read the *Racing Form*. I think it's a mark of degeneracy to say such things.

Most major newspapers publish touts' predictions for various races. Figure 18 shows the "Bay Meadows Consensus" from the *San Francisco Chronicle*. "Consensus" is the combined opinions of the experts.

I have a friend who goes skiing in the Lake Tahoe, Nevada, area. His dad is in poor health. When my friend goes skiing, he takes his dad along and drops him off in the sports book at a Lake Tahoe casino. Dad spends the day studying the *Racing Form*, betting, and socializing with casino

 PHILA. PARK 6 FURLONGS
PHILADELPHIA

6 FURLONGS. (1.08½) CLAIMING. Purse $9,000 (Plus 35% PA Bred Bonus) Fillies and Mares. 3-year-olds and upward. Weight: 3-year-olds 118 lbs. Older 122 lbs. Non-winners of two races since August 4, allowed 3 lbs. A race since August 4, 6 lbs. Claiming Price $14,000; for each $1,000 to $12,000; 2 lbs. (Races where entered for $11,000 or less not considered)

LASIX—Classic Actress, Midieval Bicker, Sweet Trade, R. Darling Daria, Cool Talc, Regal Ball, Glitz Ball, Yvanne, Strikes Nice Twice, Allen's Life.

Classic Actress
Own.—Broome E T
$14,000
B. f. 4, by Acaroid—Classic Queen, by Minnesota Mac
Br.—Kohr Elton D (Fla)
Tr.—Broome Edwin T
119

Lifetime 1990 11 2 1 4 $19,318
32 5 2 8 1989 17 2 1 3 $28,595
$58,773 Turf 8 1 1 2 $15,478

Speed Index: Last Race: +5.0 3-Race Avg.: +2.0 4-Race Avg.: −1.5 Overall Avg.: −3.9
LATEST WORKOUTS Sep 14 Pha 3f fst :37 B

Midieval Bicker
Own.—Scarlett Farm
$14,000
Ch. m. 5, by Medieval Man—Bickerbabe, by Bicker
Br.—Frank John R (Fla)
Tr.—Rowan Steve E
116

Lifetime 1990 6 0 1 1 $6,029
33 10 3 4 1989 9 2 0 1 $32,488
$124,563 Turf 2 0 0 0

Speed Index: Last Race: −2.0 3-Race Avg.: +0.6 10-Race Avg.: +2.4 Overall Avg.: +2.4
LATEST WORKOUTS Aug 17 Pha 4f fst :50½ B

Pedal Point
Own.—Cooke N
$14,000
B. m. 5, by Crafty Prospector—Slated Lady, by Slady Castle (Fla)
Br.—Cisley Stable (Fla)
Tr.—Bravo George
116

Lifetime 1990 4 0 1 1 $4,175
35 9 8 4 1989 13 3 2 2 $37,725
$111,758 Turf 2 0 1 0 $1,805

Speed Index: Last Race: −2.0 3-Race Avg.: −6.6 10-Race Avg.: −0.9 Overall Avg.: −0.9
LATEST WORKOUTS Sep 21 Pha 4f fst :51½ B

Sweet Trade
Own.—Nagel D
$12,000
B. f. 3(Apr), by El Baba—Swap Flattery, by Swaps
Br.—Cohen Zelda G (Fla)
Tr.—Velazquez Alfredo
108

Lifetime 1990 8 3 2 0 $13,350
13 5 2 0 1989 5 2 0 0 $9,600
$22,950

Speed Index: Last Race: +1.0 3-Race Avg.: −0.6 10-Race Avg.: −6.2 Overall Avg.: −6.2
LATEST WORKOUTS Aug 5 Pha 4f fst :47½ H

R. Darling Daria
Own.—Richman J D
$14,000
Ch. f. 4, by Better Arbitor—R Little Doctor, by George Lewis
Br.—Richman J D (NJ)
Tr.—Sleeter Gerald F
116

Lifetime 1990 14 0 4 3 $20,854
33 4 5 7 1989 17 4 1 4 $44,018
$64,872 Turf 2 0 0 0

Speed Index: Last Race: −7.0 3-Race Avg.: −1.6 9-Race Avg.: −0.3 Overall Avg.: −2.0
LATEST WORKOUTS Aug 9 Pha 4f fst :47½ H

FIGURE 17

San Francisco Chronicle

BAY MEADOWS CONSENSUS

	TRICKY DICK	CLOCKER	HOTSHOT	CONSENSUS
1—	Calehuche Macho Comacho Island Day Break	Island Day Break Go To The Windw Island Day Break	Macho Comacho Zar Moro Calehuche	Island Dy Brek 12 Macho Comach 11 Calehuche 7
2—Just Boogiein' By Dumas	McElhatton Born To Be King	Runing Thin Mamma's Brthdy	Mamma's Brthdy Jack Roberts McElhatton	Jack Roberts 8 McElhatton 7 Jst Boogiein' By 6
3—Chantalong	Piles Of Pleasure Pas De Bourree	Shady Speciation Pas De Bourree Rch Crem's Best	Pas De Bourree Caranomi Chantalong	Shady Speclatn 10 Pas De Bourree 10 Chantalong 9
4—Chakoti	Media Man Duplicate Light	Principal Willie Nanbob Media Man	Happy Mike Royal Bolide Cordova Cobra	Chakoti 10 Happy Mike 8 Principal Willie 5
5—Wild Pursuit	Fleet Stanley Onward Mel	Rocket Rod Fleet Stanley Agori Luck	Trv Boi Mark The Lark Rocket Rod	Fleet Stanley 11 Rocket Rod 9 Trv Boi 6
6—Hot Metal	Bom Twn Charlie Spigot	Bom Twn Charlie Our Brave Astrology	Metro Moon Astrology Trveled In Serch	Hot Metal 10 Bm Twn Charlie 8 Astrology 7
SP—Olympic Prspect	Sam Who Reconnoitering	Olympic Prspect Cutter Sam Olympic Prspect	Sam Who Sunny Blossom Sam Wno	Olympic Prspct 16 Sam Who 12 Sunny Blossom 4
7—Hatti	Keen Lady Miss Quilla IIIa	Keen Lady Storm Ryder Miss Quilla IIIa	Miss Quilla IIIa Hatti Abstract Energy	Keen Lady 13 Miss Quilla IIIa 10 Hatti 9
8—Super Mario	Kansas City Drag Race	Gold Finale Khal On The Irish Restless Con	Drag Race Captain Starbuck Kansas City	Kansas City 9 Drag Race 7 Captain Starbck 6
9—Onzas Champion	Ember's Plesre Nalee's Affair	No Money Down Onzas Champion Mikeyikeshim	Emperdori No Money Down West Jet	Onzas Champin 13 No Money Dwn 11 Emperdori 6
10—Birthday Roses	Steady Silver Glorvized	Badnewstrvelsfst Glorvized Love That Gold	Love That Gold Glorvized Badnewstrvelsfst	Love That Gold 9 Steady Silver 8 Bdnewsrvelsfst 7

Consensus (including Handicap selections) figured five points for first, three for second, one for third

STUMES' BEST BETS

■ ISLAND DAY BREAK has won three straight races while climbing back up the claiming ladder and was taken in his last two — most recently by Bryan Webb for $16,000 on December 16. The 4-year-old gelding faces his toughest test yet rising to $20,000 in today's first race but he couldn't be any sharper.

■ SHADY SPECULATION appeared on the verge of becoming one of the best local 3-year-old fillies after three straight wins against horses her own age and a solid fourth vs. older foes. She returns from a 3½-month layoff in the third, and her workouts suggest she is ready.

■ The speedy KEEN LADY opened a lead and held on well to be the fourth nose on the wire in the Charles H. Russell Handicap on November 12. She has worked in typically fast fashion since then (:58 2/5, 1:12) and doesn't have much other early speed to contend with in the seventh.

■ KANSAS CITY may prove a brilliant claim by the hot Clover Racing Stable. After being taken November 24 at Hollywood (a nine-length victory), the colt was shipped north to win an allowance race by eight. The California Juvenile Stakes represents a big step up in class, with Del Mar Futurity winner DRAG RACE and consistent D. Wayne Lukas invader CAPTAIN STARBUCK in the field.

Best Chance Plays

■ PRINCIPAL WILLIE in the fourth, BOOM TOWN CHARLIE in the sixth and BIRTHDAY ROSES in the 10th.

FIGURE 18

employees, who treat him like family. If you go to sports books in Nevada, you will see similar scenes of entertainment for senior citizens.

The major source of information for bettors who prefer not to be distracted by statistics is the program printed by every track for the day's races. Figure 19 is a page from the Bay Meadows program for December 20, 1989, showing a sample race listing with a key to the information given. The nine horses entered in the race are listed in post position order. Programs generally cost 50 cents to $1 and offer a lot of information for the money.

Tout sheets, which list the touts' picks for the day for each race, cost anywhere from $1 up, and the information they contain is typically useless. Tout services are similar to the sports services described in Chapter 5. In most cases, bettors could do as well by tacking the racing program on a wall and throwing darts to determine which horses to bet on.

OFF TRACK BETTING

Legal betting is available at off-track locations in numerous states (e.g., sports books in Nevada, OTB parlors in New York). Illegal off-track betting is also widespread.

Money bet legally off track is wired directly into the track betting pools, except in Nevada. Bets made in Nevada sports books and with illegal bookies don't become part of the pari-mutuel handle at the track; however, these betting establishments offer track odds. If sports book betting is disproportionate to track betting, books can lose. But since the 17% track take is large and since off-track betting patterns are typically close to track betting patterns, bookies and sports books usually make a profit.

LOOKING FOR THE EDGE — LEGAL OR OTHERWISE

Overlays. There are numerous so-called scientific systems and techniques for handicapping horses. Since the pari-mutuel system pits you against other gamblers, it seems reasonable to look for situations where the pari-mutuel odds don't reflect the true strengths of the horses. A wager where the odds favor the bettor is called an "overlay." For example, a claiming race where an owner enters a good horse in a low claiming

Understanding the Program

Distance of race

Arrow points to where race begins

Purse money for race

Conditions defining eligibility of horses in race

Track record for distance

Morning line odds to win as suggested by handicapper before wagering begins

Two-horse entry representing one betting interest

Horse

Owner

Trainer

Jockey

Horse's Color, Sex, Age (Bay, gelding, year foaled)

Horse's sire and dam

Color of owner's silks worn by jockey

Weight to be carried by jockey

Asterisk denotes five-pound apprentice jockey allowance

Program number also same on tote board for betting

Denotes post position in starting gate if different from program number

Denotes horse is being treated with the drug Lasix

Denotes equipment change from previous start

Boxes to write in winning horse numbers and payoffs

Name of race

Horses withdrawn from race earlier in day

Denotes Registered California Bred

$2 EXACTA WAGERING
SECOND HALF DAILY DOUBLE

| | WIN | PLACE | SHOW | NO. |

ABOUT 1,600 METERS

1 MILE TURF

The Longden Course

2ND RACE EX.

FMC CORPORATION

Purse $17,000. For four-year-olds and upward, non-winners of $3,000 other than maiden, claiming, starter or classified handicap. Four-year-olds, 120 lbs.; older, 121 lbs. Non-winners of a race other than claiming, starter or classified handicap at one mile or over since November 1 allowed 3 lbs.; such a race since October 15, 5 lbs.

Course Record—POSITION'S BEST, September 6, 1987, 1:34¾ (7) 116

➤ **ASK FOR HORSE BY PROGRAM NUMBER**

| Owner | | Trainer | Jockey |

Oak Cliff Stable — Michael C. Whittingham — Sub. Tr. G. T. Murphy — **20**
Green, tan block, green bar on tan sleeves
1 DANDY'S SECRET L ☞ 116 Thomas Chapman
P. P. 2 B.h.84, Riverman (Fr.)—Ack's Secret

Alexander, Mandato & Whittingham — Michael C. Whittingham — Sub. Tr. G. T. Murphy — **20**
Green, white "MW", white and pink bars on sleeves
1a SILVER STRIKE 116 Ron Warren, Jr.
P. P. 4 Gr.h.82, Zanthe—Silverthrill

Al Hoffman, H. F. T. — C. J. Jenda — **9/5**
Blue, white circle "H" on back, white sleeves
2 DUBLIN O'BARON L ☞ 116 Timothy Doocy
P. P. 1 B.g.84, Alla Breva—Milly Remilu

Sy Goldstein — Brett Fanning — Asst. Tr. Sandra L. Anjo — **15**
Yellow, yellow "G" in turquoise diamond on back, turquoise diamonds on sleeves
3 CO ACK L 115 Miguel Espindola
P. P. 3 Gr.g.85, April Axe—Colacka

Saud Bin Khaled — Jack Van Berg — Asst. Tr. Steve Walker — **15**
Light green and white stripes
4 ELITE REGENT L 115 Ron Hansen
P. P. 5 Ch.c.85, Vice Regent—Elile

Six-S Racing Stable — John O'Hara — **15**
Royal blue, multi-colored stripe, black "SIX-S"
5 DHALEEM L 116 Joe Steiner
P. P. 6 B.h.83, Lyphard—Patia

Gilbert & Tong — W. J. Gilbert — **6**
Salmon, black stripe on sleeves
6 FEDERALI L *111 Ken Shino
P. P. 7 B.h.84, Commissioner—Julienne (Mex.)

Stull & Wilson — Walter Greenman — Asst. Tr. David Doutrich — **9/2**
White, green and brown horses on back
7 JACK'S MUSIC ☞ 120 Jack Kaenel
P. P. 8 Ch.c.85, Debonair Roger—Mundus

Carmody & Levenson — Brett Fanning — Asst. Tr. Sandra L. Anjo — **10**
Black, black "BF" on white ball on back, black bars on white sleeves
8 BARGAIN STANDARD L 116 Russell Baze
P. P. 9 Br.h.82, Bargain Day—Flag Is Up

Alex Paszkeicz — Charles Foster — Asst. Tr. Albert Foster — **3**
Silver front, white "SCULPTURED HORSE" and horse on black back, white arrow stripe on black sleeves
9 FINALIZED L 116 Jerry Lambert
P. P. 10 Ch.g.83, Effervescing—Ladies Agreement

L—Lasix Medication. ①—First Time Lasix. ✦—First Time Off Lasix.
*5 Lbs. Apprentice Allowance.
Equipment Change: Dhaleem races without blinkers.
Declared: Soft Focus

APPROXIMATE PAY TO WIN (for $2.00 bet) IF ODDS ARE:

ODDS	PAY	ODDS	PAY	ODDS	PAY	ODDS	PAY
1-2	$3.00	6-5	$4.40	9-5	$5.60	7-2	$ 9.00
3-5	3.20	7-5	4.80	2	6.00	4	10.00
4-5	3.60	3-2	5.00	5-2	7.00	9-2	11.00
1	4.00	8-5	5.20	3	8.00	5	12.00

Pari-Mutuel means literally, a mutual wager, or betting against each other. It is similar to a stock transaction. When you buy a $2.00 ticket on a horse you are, in effect, buying one share in the race. The race track merely acts as the broker for the transaction and deducts a commission fixed by the State and shared by the State and the track. The track's takeout on every dollar remains the same, regardless of whether a favorite or longshot wins.

FIGURE 19

Saturday, December 23

DETERMINE HANDICAP
$50,000 Added
For Three-Year-Olds
One Mile and One-Sixteenth

Saturday, December 30

CALIFORNIA JUVENILE
STAKES — Grade III
$100,000 Guaranteed
For Two-Year-Olds — One Mile

EIGHTY-SECOND DAY — WEDNESDAY, DECEMBER 20, 1989

MANDATORY PICK 9 PAYOUT SUNDAY, DECEMBER 24

YOU MAY PURCHASE VOUCHERS FOR THE SELF-SERVICE MACHINES FROM ANY MUTUEL CLERK

SEE THE RACE AGAIN ON TELEVISION

Don't miss a thing! Take advantage of racing's most advanced closed circuit television system. Look around — televisions are everywhere. Each race will be shown again on video tape over the closed-circuit color screens as soon as the result is declared official.

While every effort is made to avoid mistakes in the program Bay Meadows accepts no responsibility for typographical or other errors.

SIGN UP TO BE A MEMBER OF BAY MEADOWS' MAILING LIST

Here's a great way to keep track of what's new and exciting at Bay Meadows. All members will receive periodic news updates along with valuable offers throughout the 1989-90 season.

Please present this completed form at the Courtesy Booth, located on the lower level of the Grandstand. Or send to:

Bay Meadows Mailing List
P. O. Box 5050
San Mateo, CA 94402

Name _____

Address _____

City/State _____ Zip_____

Phone _____

FIGURE 19 (continued)

class to pick up quick purse money can provide an overlay to the knowledgeable bettor. (As we have seen, it can also backfire.)

Another overlay situation can occur when there is an overwhelming favorite to win and more bettors make win bets on that horse than place or show bets. The payoff odds for a place or show bet may be better than the odds for a win bet on the same horse. Statistician David Harville did research on this and discovered that occasionally place and show bets on big favorites provide an edge to the bettor. Unfortunately, you can only spot these situations by watching the tote board, doing the arithmetic, and betting just before the windows close, a tedious task. You might as well be a stockbroker.

Steam. Occasionally, bettors have inside information on a race. Maybe a horse is sick or injured. Maybe a jockey was out partying the night before. Maybe a stable has a "good thing" they've been hiding. A bettor who has accurate inside information will bet a bundle. The pari-mutuel betting system insures that the track gets its revenue. If a huge amount of money is bet at the track, the payoff odds for a horse become lower. On the other hand, if the money is bet with a bookie, track odds are unaffected. The bookie pays track odds. Thus, someone with inside information may prefer to bet with a bookie because the bet won't adversely affect payoff odds.

For example, suppose the win pool is $9000, including a $1000 bet on Yomama to win. Without deducting track take, the payoff odds for Yomama are 8 to 1. If someone now bets $3000 on Yomama, the win pool becomes $12,000 and the payoff odds on Yomama are reduced to 2 to 1. On the other hand, if the $3000 is bet with a bookie, the track odds remain 8 to 1. If Yomama comes in first, the $3000 bettor wins $24,000 if the bet is placed with a bookie but only $6000 if the bet is placed at the track (not deducting track take). Betting with bookies to maintain track odds is called "steam." Because of steam, bookies and sports books have betting and odds limits. Known horse players are monitored, and unusually large bets are prohibited.

An extreme case of steam is known as "the build-up." Suppose you know that a particular horse, racing at a small track with light betting action, is highly likely to win. Even though this horse should be the favorite, by placing fair-sized bets at the track on all the *other* horses in the race, you can make this horse a long shot. You then bet a huge amount at a Nevada sports book or with your neighborhood bookie. When your horse wins, the bookie has to pay you the high track odds generated by your relatively small bets at the track on the other horses.

Steam will soon be obsolete at Nevada sports books. Sports book wagering will be wired directly into track betting pools. Betting at legal sports books will then be equivalent, from the bettor's point of view, to betting at the track. Only illegal bookies will have to worry about steam.

Sports books and illegal bookies suffer from betting scams. Most bettors eventually lose, however, insuring that the books make a profit. As oddsmaker Michael Roxborough puts it,

··

Always allow loyal and consistent customers their regular play regardless of the financial position of the house on the race. Regulars are the house's bread and butter so we don't want to do anything to alienate them. This doesn't mean you allow a regular $10 bettor to bet $200 on a hot horse. It means you allow a regular $10 bettor to bet $10 on a hot horse.

On-Track Betting. Not all illegal betting takes place off-track. An attorney friend of mine had a Chinese bookie for a client. In the San Francisco area there are thousands of Chinese gamblers who speak poor English. It is difficult for Chinese-speaking bettors to place bets at the track, especially in the hectic few minutes before the betting windows close. The Chinese bookie decided to alleviate this problem by setting up shop at the track. On a typical day at Golden Gate Fields in Albany, California, bettors would be lined up as usual at the betting windows. There would also be another betting line in the grandstand, where Chinese-speaking bettors lined up to place bets with *their* bookie. Needless to say, the track didn't like the competition, even though it added an international flavor to the grandstand. The Chinese bookie's on-track betting operation was short-lived.

PRISONER'S DILEMMA

I never hold a grudge. As soon as I get even with the son-of-a-bitch, I forget it.

—W. C. Fields

If you play casino games, you'll probably end up a loser, but what about other kinds of games? Gambling games—from roulette to racing—pit player against player or player against casino. There are many interactions, however, in which one might regard the participants as "players" yet in which the best course of action is not necessarily a purely competitive strategy. For example, in situations involving conflict resolution, whether the conflict is between people, businesses, or countries, it makes sense for the involved parties to consider alternatives to competitive strategies. In interactions in which there can be cooperation as well as competition, basic questions are:

When should I cooperate, and when should I refuse to cooperate?

If I decide to cooperate, how can I avoid being exploited?

In such interactions, you are not gambling for money, but gambling that if you behave a certain way, your opponent will behave a certain way. These situations can be examined in the context of a game model formulated by A. W. Tucker in 1950, called Prisoner's Dilemma.

In its original form, the game involved two prisoners accused of committing a crime who are put in isolated cells and encouraged to turn state's evidence on each other. If both remain silent, they both receive a lighter sentence than if both confess. If one cooperates with the police and the other remains silent, the one who cooperates goes free and the one who remains silent gets a stiff sentence. The "dilemma" is this: If both prisoners remain silent, they both do better than if both confess. However, without knowing what the other will do, the only way to avoid being double-crossed is to confess.

THE GENERAL VERSION

People soon realized that Prisoner's Dilemma had broad implications and provided a theoretical approach to answering the two questions cited earlier.

Here's a general version of the game. The players (opponents) have two choices: to cooperate or not cooperate. From now on I will refer to "not cooperating" as "being nasty." Each player moves simultaneously, without knowledge of the opponent's move. If they are both cooperative, they both do better than if they are both nasty. However, if one player is nasty and the other is cooperative, the nasty player double-crosses the cooperator. The choices can be represented in a display called a payoff matrix as follows, where for a given pair of moves, player A's payoff is the number on the left and player B's payoff is the number on the right:

$$
\begin{array}{c}
& & \textbf{B} \\
& & \text{Cooperate} \qquad \text{Be nasty} \\
\text{A} \quad \begin{array}{c} \text{Cooperate} \\ \text{Be nasty} \end{array} & \begin{bmatrix} (R = 3, R = 3) & (S = 0, T = 5) \\ (T = 5, S = 0) & (P = 1, P = 1) \end{bmatrix}
\end{array}
$$

and where R = "reward" for mutual cooperation

P = "punishment" for mutual nastiness

T = "temptation" payoff for taking advantage of a cooperative opponent

S = "sucker" payoff for being cooperative when opponent is nasty

If both players cooperate, they each get the reward payoff of 3 points. If both are nasty, they each get the punishment payoff of 1 point. If one player cooperates and the other is nasty, the nasty player gets 5 and the cooperator gets 0.

In general, a game is called Prisoner's Dilemma when these conditions hold: the temptation payoff is largest, followed by the reward for mutual cooperation, followed by the punishment for mutual nastiness, followed by the sucker payoff, and when mutual cooperation in repeated play is better than alternating cooperative and nasty moves with one's opponent $[R > (T + S)/2]$.

I'll start analysis with the payoffs as given above:

$$
\begin{array}{c c}
 & \begin{array}{cc} \text{Cooperate} & \text{Be nasty} \end{array} \\
\begin{array}{c} \text{Cooperate} \\ \text{Be nasty} \end{array} &
\left[\begin{array}{cc} (3,\,3) & (0,\,5) \\ (5,\,0) & (1,\,1) \end{array} \right]
\end{array}
$$

Being nasty "dominates" cooperation: No matter what one player does, it's always better for the other player to be nasty than to cooperate. If one player cooperates, the other gets 5 for being nasty, 3 for cooperating. If one player is nasty, the other gets 1 for being nasty, 0 for cooperating. The only sensible choice is to be nasty.

Again, the dilemma: Even though being nasty gives a higher payoff than cooperation, regardless of the opponent's move, if both players cooperate, they both do better than if both are nasty.

REPEATED PLAY

Suppose you and your opponent decide to play more than once. Suppose you play twice. Your score for the game will equal the sum of your payoffs from each move. You may as well be nasty on the second (and last) move since there's no future after that. Your opponent knows this, so assume your opponent will also be nasty on the second move. Since you'll both be nasty on the second move, why be nice on the first move? Be nasty on both moves!

How about playing a *random* number of moves? For example, suppose after each move, you toss a coin. If the coin comes up heads, stop. If it comes up tails, play again. Keep playing until the coin lands heads. Your score for the game is the sum of your payoffs.

Playing a random number of moves is different from playing a fixed number of moves, because there's no predetermined last move. There's always a possible future and therefore hope for mutual cooperation. A random number of moves is realistic in many situations. Whether you're bargaining with a foreign power, negotiating a business agreement, or arguing with a spouse, you don't know for sure when the game will end.

We'll analyze Prisoner's Dilemma with a random number of moves. This is called "iterated" Prisoner's Dilemma.

Let's say you play repeatedly, tossing a coin after each move and stopping the game when the coin lands heads. Your score is the sum of your payoffs. Again, here is the payoff matrix for each move.

$$
\begin{array}{c}
 & \begin{array}{cc} \text{Cooperate} & \text{Be nasty} \end{array} \\
\begin{array}{c} \text{Cooperate} \\ \text{Be nasty} \end{array} &
\begin{bmatrix} (3,\, 3) & (0,\, 5) \\ (5,\, 0) & (1,\, 1) \end{bmatrix}
\end{array}
$$

Suppose you and your opponent always cooperate. Your payoff is 3 on every move. Your total score depends on how many times you play, that is, on when the coin lands heads. For example, if heads comes up for the first time after the fourth move, the game stops and your score equals $3 + 3 + 3 + 3 = 12$. If heads comes up for the first time after the seventh move, your score equals $3 \times 7 = 21$.

Tossing a coin isn't the only way to decide when to stop. For example, you may decide to roll a pair of dice after each move and stop when the dice come up 12. Then the chance of stopping is 1/36, and the game is likely to last longer than one that is stopped as soon as a coin lands heads. The smaller the chance of stopping, the "brighter" the future. If the game is likely to last a long time, it may be possible to develop a cooperative strategy. If the game won't last long, the future isn't bright, and you might as well be nasty.

STRATEGIES FOR REPEATED PLAY

Players move with knowledge of the previous history of play. A strategy is a rule that tells you whether to cooperate or be nasty on the current move for any history of play.

Here are some popular strategies:

All C. Cooperate on every move.

All N. Be nasty on every move.

Retaliation. Cooperate as long as your opponent cooperates, but if your opponent is ever nasty, be nasty from then on.

Tit for Tat. Cooperate on the first move. Thereafter, do whatever the opponent did on the previous move: Cooperate if the opponent cooperated, be nasty if the opponent was nasty.

Random. On each play, toss a coin. If the coin lands heads, cooperate; if the coin lands tails, be nasty.

Alternate. Cooperate on every even-numbered move and be nasty on every odd-numbered move.

Fraction. Cooperate on the first move. Thereafter, cooperate if the fraction of times your opponent has cooperated until now is greater than 1/2, and be nasty if the fraction of times your opponent has cooperated is less than or equal to 1/2.

Which of these is a *good* strategy? In the one-move game, nastiness dominates cooperation. No matter what your opponent does, you do better if you're nasty than if you cooperate.

In the iterated game, whenever there is a bright future, there is no strategy that is better than *all* other strategies. All N (always be nasty) does best against All N, because any time you cooperate with All N you get the sucker payoff. On the other hand, All C (always cooperate) does best against Retaliation, because the temptation payoff you would get from a nasty move is eventually negated by Retaliation's retaliation. Thus, even the choice of whether to cooperate or be nasty on the first move depends on the opponent's strategy.

In fact, it can be shown that when there is a bright future, for *any* two distinct strategies, A and B, there is a strategy C against which A does better than B, and a strategy D against which B does better than A. Every dog has his day.

Since no strategy is always better than any other strategy, what is a good strategy? The goal is to maximize your score. If you are always nasty, you won't enjoy the benefits of playing a cooperative strategy that can't be exploited. In fact, beating your opponent can cause a decrease in your own score. On the other hand, if getting a high score means your opponent gets a higher score, so what? It's difficult to sell this idea to a competitive game player who believes that winning means doing better than one's opponent. This reminds me of a fable.

. . . .

THE SCORPION AND THE TURTLE

Once upon a time, a turtle was wandering along a river bank when a scorpion approached and said, "I can't swim. Please give me a ride across the river."

"If I give you a ride, you'll sting me," replied the turtle.

"I promise I won't sting you," said the scorpion. "I just want to get across the river."

The turtle consented, and the scorpion climbed on his back. Halfway across the river, the scorpion stung the turtle in the neck.

As the poison from the scorpion's sting took effect, the turtle started
sinking. "Why did you sting me?" moaned the turtle. "Now we'll both drown."
"I know," was the reply, "but what do you expect from a scorpion?"

In addition to nasty strategies like All N, there are "nice" strategies. A nice strategy is one that is never the first to be nasty. Always cooperating (All C), Retaliation, Tit for Tat, and Fraction are nice. All N and Random aren't nice. All C does well against any nice strategy, because a nice strategy will always cooperate with All C. Unfortunately, All C can be exploited by a nasty strategy like All N.

A good strategy should be cooperative but not allow itself to be exploited.

PRISONER'S DILEMMA PLAYED BY COMPUTERS

If you can't prove something theoretically, try a computer. In the early 1980s, Robert Axelrod, a political science professor at the University of Michigan, held two Prisoner's Dilemma tournaments to see which strategies would do well in actual play. Strategies were submitted as computer programs that were pitted against each other in a round robin tournament. Each strategy played every other strategy. The winner was the strategy with the highest score.

Fifteen strategies were in the first tournament. Each matchup consisted of five games of 200 moves, with the score for each strategy being the strategy's average score for the five games.

The winner of the first tournament was Tit for Tat, submitted by Anatol Rapoport, a psychology professor at the University of Toronto. Tit for Tat cooperates on the first move. Thereafter, it does whatever the opponent did on the previous move. It cooperates if the opponent cooperates and is nasty if the opponent is nasty.

Some of the strategies in the tournament were based on the opponent's history of play. Half the strategies were nice and half were not nice. All the nice strategies finished ahead of all the not-nice strategies.

Axelrod advertised the second tournament more widely than the first. There were 62 entries. Every contestant in the second tournament was given the results of the first tournament. Players in the second tournament were gunning for Tit for Tat, winner of the first tournament.

Tit for Tat won again!

TIT FOR TAT

. . . .

In an ancient land, there was a wise man who dressed like a beggar. Once, after a desert trek, the wise man went to a bath house. The arrogant bath house attendant was disgusted by the traveler and led him to a moldy basement shower. After the traveler showered, the attendant gave him a dirty towel. The wise man thanked the attendant and gave him a gold coin. The attendant was astonished that a beggar would have a large gold coin, let alone give it to him.

Next week, the wise man returned to the bath house. This time, the attendant treated him like royalty, escorting him to a large, sunny shower stall, giving him a stack of clean towels, powder, the works. Afterward, the wise man thanked the attendant and gave him a penny.

"What's this?!" said the enraged attendant. "Last week, when I treated you like a beggar, you gave me a gold coin. Now when I treat you like royalty, you give me a penny!"

The wise man replied, "The gold coin was for today. The penny is for last week."

Why is Tit for Tat a good strategy? It is cooperative, yet can't be exploited. It will change its behavior according to the opponent's behavior. Therefore a nasty strategy must itself change to get Tit for Tat to be nice. Tit for Tat can never get a higher score than a particular opponent, yet it won both tournaments. In the tournaments, it did well against nice strategies, wasn't exploited by nasty ones, and thus got the highest total score.

Here is the main mathematical result in Prisoner's Dilemma: When the future is bright, if you are ever nasty against Tit for Tat you will do worse than by always cooperating. If you "beat" Tit for Tat, you pull down your own score in the process.

Tit for Tat won't take advantage of another strategy. Tit for Tat and its opponent both do best when both strategies cooperate. You can trust Tit for Tat.

There are two reasons for a player to be unilaterally nasty:

1 – To take advantage of an opponent

2 – To avoid being taken advantage of by an opponent

Tit for Tat eliminates the reasons for being nasty.

Tit for Tat has other good features. As Axelrod puts it in his book *The Evolution of Cooperation*:

> What accounts for Tit for Tat's robust success is its combination of being nice, retaliatory, forgiving, and clear. Its niceness prevents it from getting into unnecessary trouble. Its retaliation discourages the other side from persisting whenever defection [nastiness] is tried. Its forgiveness helps restore mutual cooperation. And its clarity makes it intelligible to the other player, thereby eliciting long-term cooperation.

Axelrod also used Prisoner's Dilemma strategies in a model simulating "survival of the fittest." The computer generated a sequence of Prisoner's Dilemma tournaments. After each tournament, the strategies were rewarded by having copies of themselves added to the strategy population in proportion to their score. Strategies that did well in a tournament "reproduced" more than strategies that did poorly.

As cooperative strategies increased in number, they had more interactions with themselves and, because of mutual cooperation, kept growing. In the beginning, nasty strategies did well, but only at the expense of sucker strategies that could be exploited. As time passed, the strategies exploited by the nasty strategies became extinct. Eventually, the nasty strategies had no one left to exploit, so they died too. As Axelrod puts it,

> At first, poor programs and good programs are represented in equal proportions. But as time passes, the poorer ones begin to drop out and the good ones thrive. Success breeds more success, provided that the success derives from interaction with other successful rules. If, on the other hand, a decision rule's success derives from its ability to exploit other rules, then as these exploited rules die out, the exploiter's base of support becomes eroded and the exploiter suffers a similar fate. . . . Not being nice may look promising at first, but in the long run it can destroy the very environment it needs for its own success.

After 1000 "generations," Tit for Tat was the top strategy and growing at a faster rate than any other strategy. In this case, survival of the fittest did not mean survival of the nastiest. Because they wouldn't cooperate, the nasty strategies forced first their victims and then themselves into extinction.

YOU CAN WIN

Winning doesn't have to mean being nasty. It also doesn't have to mean that you got lucky in the lottery. There are many games in everyday life where you can use a dynamic strategy like Tit for Tat and where, although you may have to gamble, everyone can win.

GLOSSARY OF GAMBLING TERMS

Action. The total wagered by all bettors on a particular bet.

Bankroll. The money a gambler uses for betting.

Beard. Someone who places bets for a gambler who wants to remain anonymous. A runner.

Book. An establishment where you can bet on sports and racing.

Bookie. Someone who runs an illegal book.

Card counting. Part of a blackjack strategy in which the player keeps track of the composition of cards remaining in the deck, betting more when the deck is high in 10-value cards than when the deck is low in 10-value cards.

Chalk. Favorite. A "chalk bettor" bets on favorites.

Churn. Rebetting winnings until you are a loser.

Closing line. The final point spread before the game is played.

Covering the spread. A team "covers" the point spread if its score is higher than the opponent's score after the point spread has been subtracted from the favorite's score.

Deal seconds. A cheating technique in which the dealer deals the second card instead of the top card in the deck until the top card can be used to the dealer's advantage.

Dime. $1000.

Dog. The underdog, or "inferior," team in a sports event.

Dollar. $100.

Double-up strategy. Doubling your bet when you lose, to "cover your losses."

Dutch book. A set of odds for which the bettor is guaranteed to profit if betting is made on each event in the proper proportions.

Expected winnings. A player's long-run average winnings (or losses) in a gambling game.

Eye in the sky. One-way mirrors on the ceilings of a casino, used by the casino for surveillance of players and employees.

Fair odds. Payoff odds for which a player's expected winnings equal zero. Ratio of losses to wins. For example, if $P(\text{win}) = 1/5$, then $P(\text{lose}) = 4/5$, and fair odds equal 4 to 1. If $P(\text{win}) = 3/7$, fair odds equal 4 to 3.

Favorite. The team or horse most likely to win.

Fish. Sucker.

Futures bet. A bet on which team will win a future sports event, usually a championship game (like the Superbowl).

Handicapper. A sports or horse racing bettor who tries to predict results of games or races by studying data.

Handle. Action.

High roller. A big bettor.

House. The casino, state, track, or whoever is running the game.

House edge. The expected winnings of the house per dollar bet for a particular wager, expressed as a percentage.

Jackpot. The big prize.

Juice. The book's profit. Also known as "vig."

Law of averages. An important result in probability theory that states that in repeated play of a game under identical conditions, the observed fraction of wins approaches the theoretical probability of winning.

Law of large numbers. Law of averages.

Lay odds. To make a wager with payoff odds less than even, that is, where your winnings for a $1 bet are less than $1.

Line. The point spread or payoff odds on a sports event.

Linemaker. Oddsmaker.

Long shot. A team or horse that is unlikely to win.

Middle. A situation in which the point spread is changed to balance betting action and the difference in score lands between the point spreads, causing the book to lose both ways.

Minus pool. A "minus pool" exists when so much is bet on one horse that if the usual track take were subtracted from the betting pool, there would not be enough left to give the winning bettors a profit.

Money line. A wager on the outcome of a game in which there are odds but no point spread.

Nickel. $500.

Number. Point spread.

Odds. Payoff odds. The amount you win for a successful bet of a certain amount. For example, 3 to 1 odds means that if you bet 1 dollar and win, you get back 3 dollars plus the dollar you bet. X to Y odds means that if you bet Y dollars and win, you get paid X dollars plus the Y dollars you bet.

Oddsmaker. A linemaker (the person who determines odds and point spreads).

Opening line. The first public point spreads for a set of games.

Overlay. A wager in which the bettor has an edge.

Over-under. A bet on whether the combined score of both teams in a game will be over or under a specified number.

Pari-mutuel. A betting structure in which losers pay winners after the house takes a fixed percentage of the betting action.

Parlay. A bet on more than one game or proposition for which all specified events have to occur for the bet to be a winner.

Parlay card. A list of games and propositions on a betting card. You bet on a set of events on the card, usually from three to ten. You win if all of the events you pick win.

Past posting. Betting on a race after it has been run.

Payoff odds. See "odds."

Pick. An even game. In point-spread betting, point spread equals zero.

Player. A bettor.

Point spread. The number of points by which the favorite must beat the underdog for the bettor on the favorite to win. The purpose of a point spread is to equalize betting action on both teams.

Probability distribution. A list of probabilities associated with the numerical outcomes of a random experiment.

Proposition. A bet.

Push. A tie.

Runner. A beard.

Sharp. A wise guy.

Sports service. A company that sells information and picks to sports bettors.

Steam. Money bet with a bookie, perhaps on a "fixed" race, in order not to adversely affect track payoff odds.

Strategy. A rule that tells a player what to do throughout the course of a game.

Sucker. Someone who can be conned. A fish.

Teaser. A parlay with better point spreads but lower payoff odds than an ordinary parlay.

Take odds. Bet on the underdog. Also known as "take points."

Tit for Tat. A strategy in which you are nice whenever your opponent is nice and nasty whenever your opponent is nasty.

Tote board. A computerized display at the track that shows the amount bet on each horse in the different betting pools.

Tout. Someone who sells tips on who will win a game or race.

Track take. The percentage taken from the betting pool by the track, state and local governments before the winners are paid.

Underdog. The team most likely to lose.

Vig. The bookmaker's profit. The vigorish.

Vigorish. The book's profit. The juice.

Wise guy. A professional bettor.

BIBLIOGRAPHY

Ainslie, Tom. *Ainslie's Complete Guide to Thoroughbred Racing*, Simon and Schuster, 1986.

Ankeny, Nesmith. *Poker Strategy*, Perigee, 1981.

Axelrod, Robert. *The Evolution of Cooperation*, Basic Books, 1984.

Banker, Lem, and Frederick Klein. *Lem Banker's Book of Sports Betting*, E. P. Dutton, 1986.

Beckett, Samuel. *Waiting for Godot*, Grove Press, 1954.

Beckett, Samuel. *Endgame*, Grove Press, 1958.

Davidowitz, S. *Betting Thoroughbreds*, E. P. Dutton, 1977.

Davis, Morton. *Game Theory*, Basic Books, 1970.

Dostoyevski, Fyodor. *The Gambler*, Penguin Books, 1966.

Dubins, Lester, and Leonard J. Savage. *How to Gamble If You Must*, McGraw-Hill, 1965.

Fabricand, Burton. *The Science of Winning*, Whitlock Press, 1979.

Ferguson, Thomas. *Mathematical Statistics: A Decision Theoretic Approach*, Academic Press, 1967.

Graham, Virginia, and C. Ionescu Tulcea. *A Book on Casino Gambling*, Van Nostrand Reinhold, 1978.

Hawking, Stephen W. *A Brief History of Time*, Bantam Books, 1988.

Hoffmann, Lawrence, and Michael Orkin. *Finite Mathematics with Applications*, McGraw-Hill, 1979.

Humble, Lance, and Carl Cooper. *The World's Greatest Blackjack Book*, Doubleday, 1980.

Kafka, Franz. *Parables and Paradoxes*, Schocken Books, 1935.

Lang, Arne. "Stardust Sports Registry," January 1990.

Moldea, Dan. *Interference*, William Morrow, 1989.

Olshan, Mort. *The Best of the Gold Sheet*, Nationwide Sports Publications, 1988.

Olshan, Mort. *Winning Theories of Sports Handicapping*, Simon and Schuster, 1975.

Packel, Edward. *The Mathematics of Games and Gambling*, Mathematical Association of America, 1981.

Pinter, Harold. *The Caretaker*, Grove Press, 1960.

Roxborough, Michael, and Mike Rhoden. *Race and Sports Book Management*, Las Vegas Sports Consultants, 1989.

Savage, Leonard J. *The Foundations of Statistics*, Wiley, 1954.

Scarne, John. *Scarne's Complete Guide to Gambling*, Simon and Schuster, 1961.

Thorp, Edward. *Beat the Dealer*, Vintage Books, 1962.

Thorp, Edward. *The Mathematics of Gambling*, Gambling Times, 1984.

Tulcea, C. Ionescu. *A Book on Casino Craps*, Van Nostrand Reinhold, 1981.

Williams, John D. *The Compleat Strategyst*, McGraw-Hill, 1954.

Ziemba, William T., and Donald B. Hausch. *Dr. Z's Beat the Racetrack*, William Morrow, 1987.

FIGURING THE ODDS

CHAPTER 2 • Roulette

1. Suppose you could make a roulette bet that a number from 1 through 8 comes up and get payoff odds of 3 to 1. Find:

   a. P(you win)

   b. P(you lose)

   c. Expected winnings for a $1 bet

   d. Expected winnings for a $5 bet

   e. House edge

2. Suppose you bet $1 on 9 (35 to 1 payoff odds), $1 on 1st 12 (2 to 1 payoff odds), and $1 on 1–18 (1 to 1 payoff odds). Find:

   a. Expected winnings for the combined bet

   b. House edge for the combined bet

3. Suppose you could bet that one of the numbers "1 through 13" comes up and get 3 to 2 payoff odds. Find:

   a. P(you win)

   b. P(you lose)

   c. Expected winnings for a $2 bet

   d. House edge

4. Find fair odds for the following bets:

   a. Bet on black

   b. Bet on green (0 or 00)

   c. Bet on 1st column

   d. Bet on four numbers

   e. Bet on two numbers

5. Suppose you bet $3 on 1st 12 (payoff odds 2 to 1), $2 on 1 to 18 (payoff odds 1 to 1), and $1 on 00. Find:

   a. Expected winnings for the combined bet

   b. House edge for the combined bet

6. Suppose you bet $1 on the four numbers bet 22 23 25 26 (payoff odds 8 to 1), $1 on the first column (payoff odds 2 to 1), and $1 on 2nd 12 (payoff odds 2 to 1). Find:

   a. Expected winnings for the combined bet

   b. House edge

7. In European casinos, a roulette wheel has only 37 sections. There is no 00. There are a few similar wheels in Nevada. Payoff odds are the same

as for roulette played with 38 section wheels. Find the probability of winning, expected winnings, and house edge for the following bets using a 37-section roulette wheel:

a. Bet on red

b. Bet on 1st 12

c. Bet on 17

d. Two numbers bet

8. Suppose you make two successive red bets. Using the multiplication rule for independent events, find

a. P(you lose both bets)

b. P(you win one bet and lose one bet in some order)

9. Suppose you make three successive red bets. Using the multiplication rule for independent events, find:

a. P(you win all three bets)

b. P(you lose all three bets)

c. P(you win two bets and lose one, in some order)

d. P(you win one bet and lose two)

10. Suppose you make four successive red bets. Using the multiplication rule for independent events, find:

a. P(you win all four bets)

b. P(you lose all four bets)

c. P(you win three bets and lose one, in some order)

d. P(you win two bets and lose two)

e. P(you win one bet and lose three)

f. P(you win at least one bet)

11. You receive a notice in the mail that you may have won the Super Sucker Sweepstakes prize of $1 million. All you have to do is fill out the acceptance form and return it in the enclosed envelope. The notice goes on to say that 10 million names were entered in the drawing for the grand prize. No entry fee is necessary. You fill out the form and return it.

a. Find your expected winnings and house edge, assuming that you have to pay 25 cents for a postage stamp to send in the form.

b. Suppose the enclosed envelope has postage paid. Find your expected winnings and house edge, assuming that it takes you five minutes to fill out the form and that your time is worth $6 an hour.

12. Suppose you find a roulette wheel that is biased and for which P(red comes up) = 21/38. Find your expected winnings for red bets.

13. Suppose you find a roulette wheel that is biased and for which P (#17 comes up) = 2/38. Find your expected winnings for bets on 17.

14. Suppose you find a roulette game where, in order to attract business, the casino is offering payoff odds of 38 to 1 for bets on individual numbers. Other payoff odds are the same as ordinary roulette payoffs. The wheel is evenly balanced.

a. Find your expected winnings for bets on an individual number.

b. How could you bet with these payoff odds to guarantee winning every time the wheel was spun?

15. An insurance company sells an injury insurance policy for $20. If you get injured, you get $1000. Otherwise, you get nothing. There is a .01 chance that you get injured.

a. Find the insurance company's "edge" (expected profit).

b. How much does the insurance company have to charge for the policy in order to break even (assuming lots of customers)?

CHAPTER 3 • Craps

1. A pair of dice are rolled and the upturned dots are counted. Find the following probabilities:

a. $P(9)$

b. $P(11)$

c. $P(9 \text{ or } 11)$

d. $P(\text{odd number})$

e. $P(\text{number greater than } 7)$

f. $P(\text{number less than } 4)$

2. A pair of dice are rolled repeatedly. Find:

a. $P(5 \text{ comes up before } 7)$

b. $P(6 \text{ comes up before } 7)$

c. $P(2 \text{ comes up before } 7)$

d. $P(7 \text{ comes up before } 12)$

e. $P(\text{even number comes up before } 7)$

f. $P(5, 6, 8, \text{ or } 9 \text{ comes up before } 7)$

g. $P(8 \text{ comes up before } 9)$

h. $P(\text{even number comes up before odd number})$

3. A pair of dice are rolled. If 4 or 10 comes up, you win. Otherwise, you lose. Payoff odds are 4 to 1. Find:

a. $P(\text{you win})$

b. $P(\text{you lose})$

c. Expected winnings for a $1 bet

d. House edge

e. Fair odds

4. A pair of dice are rolled. If 8, 9, or 10 come up, you win. Otherwise you lose. Payoff odds are 2 to 1. Find:

a. $P(\text{you win})$

b. $P(\text{you lose})$

    c. Expected winnings for a $1 bet

    d. House edge

    e. Fair odds

5. Someone offers you the following bet. A pair of dice are rolled. If 6, 7, or 8 come up, you win. Otherwise you lose. Payoff odds are 3 to 2. Find:

    a. *P*(you win)

    b. *P*(you lose)

    c. Expected winnings for a $2 bet

    d. Expected winnings per $1 bet

    e. Should you make this bet?

6. You make one $10 pass line bet. What happens to your bet for the following sequences of rolls (the first roll is a come-out roll)?

    a. 5, 6, 9, 5, . . .   f. 3, 7, . . .
    b. 9, 8, 11, 3, 2, 12, 9, . . .   g. 11, 5, 8, . . .
    c. 4, 6, 11, 9, 2, 7, . . .   h. 8, 4, 2, 3, 11, 12, 9, 5, 6, . . .
    d. 8, 5, 7, 8, . . .   i. 2, 3, 7, 11, 12, . . .
    e. 7, 3, . . .   j. 12, 6, 6, 6, 7, . . .

7. You make one $20 pass line bet. If a point is established, you then make a $40 odds bet. How much do you win or lose for the following sequences of rolls?

    a. 6, 8, 9, 11, 3, 6, . . .   f. 12, 9, 9, . . .
    b. 9, 8, 3, 2, 9, . . .   g. 11, 4, 4, 8, 7, . . .
    c. 10, 5, 11, 12, 10, . . .   h. 3, 2, 12, 7, 10, . . .
    d. 5, 9, 6, 7, . . .   i. 2, 2, 2, 2, . . .
    e. 7, 3, 4, 4, . . .

8. You make one $25 come bet. What happens to your bet for the following sequences of rolls?

    a. 8, 9, 10, 9, 3, 12, 7, 8, . . .

    b. 6, 3, 2, 12, 9, 8, 7, . . .

    c. 7, 11, 2, 11, 9, . . .

    d. 11, 7, 7, 11, . . .

    e. 12, 5, 5, 5, 5, . . .

    f. 2, 8, 3, 7, 8, . . .

    g. 3, 9, 6, 9, 7, . . .

9. You make one $30 don't come bet. What happens to your bet for the sequences of rolls in Problem 8?

10. You bet $10 that the next roll of the dice is 7 (payoff odds are 4 to 1).

    a. The next roll is 7. How much do you win?

    b. You make this bet twice. Find *P*(win both).

    c. You make this bet three times. Find *P*(win all three).

11. You make a $40 field bet. How much do you win or lose when the next roll of the dice is:

a. 7

b. 2

c. 11

12. You make one $20 place bet on 4. How much do you win or lose for the following sequences of rolls?

a. 5, 6, 9, 11, 12, 2, 5, 4, 8, 3, . . .

b. 7, 3, 6, 4, . . .

c. 9, 8, 11, 3, 2, 7, 4, . . .

13. You make one $30 place bet on 8. How much do you win or lose for the following sequences of rolls?

a. 7, 4, 3, 12, 8, . . .

b. 6, 9, 6, 7, 5, 8, . . .

c. 12, 3, 2, 8, 5, 10, . . .

14. You make one $60 place bet on 6. How much do you win or lose for the following sequences of rolls?

a. 5, 8, 9, 3, 2, 6, . . .

b. 9, 8, 2, 3, 9, 4, 7, 6, . . .

c. 6, 3, 7, . . .

15. You make one $60 big 6 bet. How much do you win or lose for the sequences of rolls in Problem 14?

16. You make one $30 hardway 6 bet. How much do you win or lose for the following sequences of rolls?

a. 4 5, 5 6, 1 1, 4 4, 5 2, . . .

b. 3 2, 3 1, 5 5, 2 4, 3 6, 3 3, . . .

c. 5 6, 4 4, 1 5, 3 3, 5 1, 5 2, . . .

17. You make one $50 hardway 10 bet. How much do you win or lose for the following sequences of rolls?

a. 4 6, 5 4, 6 2, . . .

b. 5 1, 5 1, 3 3, 6 2, 5 5, 1 5, . . .

c. 4 4, 3 1, 4 5, 6 2, 6 1, . . .

18. Find your average total winnings or losses if you make 1000 $5 bets on each of the following bets.

a. Pass line   g. Any craps bet
b. Don't pass   h. Hardway 8 bet
c. Come   i. Hardway 4 bet
d. Don't come   j. Pass line bet followed by $5 odds
e. Place bet on 6   bet whenever possible
f. Place bet on 10

19. At Vegas World Casino in Las Vegas, you can play a version of craps known as "crapless craps," in which, for a pass line bet, every number except 7 is a point: If 7 is rolled on the come-out roll, you win. Otherwise, the come-out roll (2, 3, 4, 5, 6, 8, 9, 10, 11, or 12) is a point. As in

ordinary craps, you win if the point comes up before 7, and lose if 7 comes up before the point. Payoff odds are 1 to 1, same as in ordinary craps. Suppose you make a pass line bet in crapless craps. Find:

a. P(you win)

b. P(you lose)

c. House edge

20. If you make a pass line bet in crapless craps, you can make an odds bet when a point is established of up to 12 times the pass line bet. As in ordinary craps, you get fair payoff odds for the odds bet. Find payoff odds for odds bets on points 2, 3, 11, and 12.

21. Suppose you make $5 pass line bets in crapless craps, backing them up with $60 odds bets whenever you can. How much do you win or lose for the following sequences of rolls?

a. 5, 6, 4, 9, 2, 3, 12, 5, . . .

b. 9, 11, 12, 2, 3, 7, . . .

c. 2, 5, 10, 12, 9, 6, 2, . . .

d. 11, 3, 4, 6, 8, 9, 7, 3, . . .

e. 12, 6, 12, 6, 7, . . .

f. 7, 4, 5, 12, 4, . . .

g. 3, 12, 9, 10, 3, 7, . . .

22. In the carnival game of chuck-a-luck, three dice are rolled. You bet on one of the numbers from 1 to 6. You win the amount you bet every time your number comes up. For example, if you make a $1 bet on number 1, you win $1 for each 1 that comes up: If 1 comes up on all three dice, you win $3. If 1 comes up on two of the three dice, you win $2. If 1 comes up on one of the three dice, you win $1. If 1 doesn't come up on any of the dice, you lose your $1 bet. For a $1 bet, find:

a. P(you win $3)

b. P(you win $2)

c. P(you win $1)

d. P(you lose $1)

e. Expected winnings

f. House edge

23. In Chapter 1, I discussed the French gambler the Chevalier de Mère and his two dice games:

> Game 1—A die is rolled four times. You bet that no 6s will come up. Payoff odds are 1 to 1.
>
> Game 2—A pair of dice are rolled 24 times. You bet that no double-6s will come up. Payoff odds are 1 to 1.

Find P(you win) and your expected winnings for a $1 bet for each of the Chevalier's two games.

CHAPTER 4 • Keno and Slots

1. You make a $2, 10-number keno bet. You match 7 numbers, that is, 7 of your numbers are among the 20 selected. According to the keno payoff table on page 47, how much is your payoff?

2. You make a $2, 10-number keno bet on the numbers 3, 6, 9, 12, 24, 36, 48, 60, 70, and 80. The winning numbers are 3, 9, 17, 19, 35, 36, 39, 41, 47, 54, 58, 60, 64, 65, 67, 70, 71, 75, 77, and 79. How much is your payoff?

3. You make a $2, 10-number keno bet. You match four numbers. According to the payoff table, how much is your payoff?

4. You make a $50, 10-number keno bet. You match eight numbers. According to the payoff table, how much is your payoff?

5. In some casinos you can make a keno bet on one number. If your number is matched, a $2 ticket pays $6, yielding a $4 profit. Since 20 numbers are randomly selected from 80, the chance that your number is selected is $20/80 = 1/4$ [equivalently, $\binom{1}{1}\binom{79}{19}/\binom{80}{20}$]. Find the house edge for the 1-number bet.

6. If you make a $2, 5-number keno bet, payoffs and probabilities are as follows:

Match	Payoff	Probability
3	$ 2	.0839
4	20	.0121
5	1500	.0006

Find:

a. Expected payoff
b. Expected winnings (winnings = payoff minus cost of ticket)
c. House edge
d. P(you lose $2)

7. How many permutations are there of the numbers 1 2 3 4 5 6 7?

8. How may ways can 11 people be arranged in a line?

9. Compute the following:

a. $10!$

b. $10!/(7! \times 3!)$

c. $\binom{10}{7}$

d. $\binom{10}{3}$

e. $\binom{10}{10}$

10. How many distinguishable arrangements are there of the letters A A A A A B B B B C C C D D E?

11. How many distinguishable arrangements are there of the numbers 1 1 1 1 1 2 2 2 2 3 3 3 4 4 5?

12. How many distinguishable arrangements are there of the numbers 1 1 1 1 1 1 1 0 0 0 0?

13. How many ways are there to select 7 people from a group of 11 people?

14. How many ways are there to select 4 people from a group of 11 people?

15. How many ways are there to select 9 frogs from a group of 17 frogs?

16. How many ways are there to select 17 frogs from a group of 17 frogs?

17. How many ways are there to select k frogs from a group of n frogs $(0 \leq k \leq n)$?

18. How many ways are there to select $n - k$ frogs from a group of n frogs?

19. Nine of thirty applicants for the President's Advisory Panel on Gambling are professional gamblers. Seven applicants are selected at random to be on the panel. Find the probability that three of the selected applicants are gamblers (give the formula only).

20. Five cards are dealt from an ordinary deck of cards (as in poker). Using the combinations formula, find the correct expression for (assume random selection):

   a. P(all hearts)

   b. P(all of same suit)

   c. P(three queens, two kings)

   d. P(three of one denomination, two of another) i.e., P(full house)

   e. P(three of one denomination, other two of different denominations) i.e., P(three of a kind)

   f. P(two of one denomination, other three of different denominations) i.e., P(one pair)

21. A box contains 20 tickets, 5 of which are "winners." Three tickets are selected at random. Find (formulas only):

   a. P(all winners)

   b. P(two winners)

   c. P(one winner)

   d. P(no winners)

   e. P(at least one winner)

   f. Expected number of winners (*Hint:* The chance of one selected ticket being a winner equals 1/4.)

22. You make a keno bet on 7 numbers. Find (formulas only):

   a. P(you match 4)

   b. P(you match 5)

   c. P(you match 6)

d. P(you match 7)

23. You make a keno bet on 12 numbers. Find (formulas only):

   a. P(you match 6)

   b. P(you match 7)

   c. P(you match 8)

   d. P(you match 9)

   e. P(you match 12)

   f. P(you match none)

24. In some casinos, there is a special keno bet in which you can bet on 20 numbers, with a $5 minimum bet. If you have 7 or more matches, you get a payoff. If you have 4, 5, or 6 matches, you lose your bet. However, there is a payoff for 1, 2, or 3 matches. In fact, for a $5 bet, if *none* of your numbers are matched, you are paid $600! Find (formulas only):

   a. P(none of your numbers are matched)

   b. P(5 of your numbers are matched)

   c. P(10 of your numbers are matched)

   d. P(all 20 of your numbers are matched)

   e. Expected number of matches (*Hint*: Since there is a 25% chance that any particular number will be matched, as in Problem 5, on the average 1 out of every 4 numbers will be matched.)

25. A quarter slot machine has three reels, each with 10 symbols. Reel 1 has three grapefruits and two cherries. Reel 2 has two grapefruits and one cherry. Reel 3 has one grapefruit and three cherries. Assume that the reels spin independently and that every symbol has the same chance of coming up. You put in a quarter and pull the handle. Find:

   a. P(three cherries)

   b. P(two cherries and a grapefruit)

   c. P(two grapefruits and a cherry)

   d. P(three grapefruits)

26. A $1 slot machine is advertised to pay back "95%." What is your hourly loss if you put $30 per minute into the machine?

CHAPTER 5 • Sports Betting

1. The Los Angeles Raiders are playing the Seattle Seahawks. The line is:

                    Raiders
                    Seahawks   6

   The Seahawks win, 28–24.

   a. Which team is the favorite?

   b. Which team is the home team?

   c. What must happen for Seahawks bettors to win?

   d. You bet $22 on the Raiders. What happens to your bet?

e. You bet $33 on the Seahawks. What happens to your bet?

2. The Washington Redskins are playing the New York Giants.

Redskins   43
Giants   3½

The Giants win, 34–17.

   a. Which team is the underdog?

   b. Which team is the favorite?

   c. You bet $55 on the Giants. What happens to your bet?

   d. You bet $77 on the Redskins. What happens to your bet?

   e. You bet $110 on "over." What happens to your bet?

   f. You bet $220 on "under." What happens to your bet?

3. The Denver Broncos are playing the Green Bay Packers.

Broncos   10
Packers   42

The game is a tie, 17–17.

   a. Which team is the underdog?

   b. Which team is the favorite?

   c. You bet $550 on the Broncos. What happens to your bet?

   d. You bet $77 on the Packers. What happens to your bet?

   e. You bet $1100 on "over." What happens to your bet?

   f. You bet $880 on "under." What happens to your bet?

4. The Miami Dolphins are playing the New York Jets:

Dolphins   3
Jets   47½

The Jets win, 27–25.

   a. Which team is the favorite?

   b. You bet $100 on the Dolphins. What happens to your bet?

   c. You bet $50 on the Jets. What happens to your bet?

   d. You bet $110 on "over." What happens to your bet?

5. The Buffalo Bills are playing the Indianapolis Colts:

Colts   40
Bills   6

The Bills win, 23–17.

   a. Which team is the underdog?

   b. You bet $440 on the Bills. What happens to your bet?

   c. You bet $550 on the Colts. What happens to your bet?

   d. You bet $660 on "under." What happens to your bet?

   e. You bet $770 on "over." What happens to your bet?

6. The Los Angeles Rams are playing the Atlanta Falcons. The opening line makes the Rams a 5-point favorite. You bet $110 on the Falcons and take the 5 points. More money is bet on the Falcons than the Rams. The point spread is moved to make the Rams a 3-point favorite. Now you bet $110 on the Rams and give 3 points. You have thus bet $110 on each team.

    a. The Rams win, 35–21. What happens to your bets?

    b. The Rams win, 25–21. What happens to your bets?

    c. The Rams win, 35–30. What happens to your bets?

    d. The Rams win, 37–34. What happens to your bets?

    e. The Falcons win, 21–20. What happens to your bets?

    f. The Rams win, 25–21. Most of the bets on the Rams were at Rams −3, and most of the bets on the Falcons were at Rams −5. What happens to the sports books?

7. The Kansas City Chiefs are playing the Los Angeles Raiders:

    $$\text{Raiders} \qquad 5\frac{1}{2}$$
    $$\text{Chiefs}$$

    You think $P(\text{Chiefs cover spread}) = .65$. You bet $220 on the Chiefs. Assuming the probability is correct, find your expected winnings and expected winnings per $1 bet.

8. Suppose your assumption in Problem 7 is wrong and actually $P(\text{Chiefs cover spread}) = .51$. You bet $220 on the Chiefs. Find your expected winnings and expected winnings per $1 bet.

9. The Detroit Lions are playing the Tampa Bay Buccaneers. Detroit is a 3-point favorite. At a certain sports book, $75,000 is bet on Detroit and $75,000 is bet on Tampa Bay.

    a. What is the sports book's profit per $1 bet if Tampa Bay covers the spread?

    b. What is the sports book's profit per $1 bet if Detroit covers the spread?

    c. What is the sports book's profit per $1 bet if Detroit wins 24–21?

10. The Cincinnati Bengals are playing the New York Jets. You think $P(\text{Cincinnati covers spread}) = .60$. You have $500. Using the Kelly system, how much should you bet?

11. The New England Patriots are playing the Miami Dolphins. You think $P(\text{New England covers spread}) = .58$. You have $1000. Using the Kelly system, how much should you bet?

12. The Philadelphia Eagles are playing the Washington Redskins. You think $P(\text{Redskins cover spread}) = .67$. You have $560. Using the Kelly system, how much should you bet?

13. The San Francisco 49ers are playing the Los Angeles Rams. You think $P(\text{49ers cover spread}) = .51$. You have $700. Using the Kelly system, how much should you bet?

For Problems 14 through 17, assume that your picks cover the spread 55% of the time.

14. You start the season with a bankroll of $5500. You decide to bet your entire bankroll on each bet.

    a. Find your expected winnings for the first bet.

    b. Find the probability that you will be broke after 10 bets. (*Hint*: Since you bet your entire bankroll on each bet, you go broke as soon as you lose.)

15. You start the season with $5500. You decide to bet $55 on every bet.

    a. Find your expected winnings for the first bet.

    b. Find the probability that you will be broke after 10 bets.

    c. Find your expected winnings after 10 bets.

    d. Find your expected winnings after 100 bets.

16. You start the season with $5500. You decide to bet 40% of your current bankroll on each bet.

    a. Find your expected winnings for the first bet.

    b. Find the probability that you will be broke after 10 bets.

17. You start the season with $5500. You use the Kelly system.

    a. Find your expected winnings for your first bet.

    b. Find the probability that you will be broke after 10 bets.

18. The Cowboys are 5-point favorites over the Lions. The Raiders are 6½-point favorites over the Chiefs. You make a $50 parlay bet on the Cowboys and the Chiefs.

    a. The Cowboys win, 35–14, and the Raiders win, 21–17. What happens to your bet?

    b. The Cowboys win, 35–31, and the Raiders win, 21–17. What happens to your bet?

    c. The Cowboys win, 21–7, and the Chiefs win, 21–20. What happens to your bet?

    d. The Lions win, 14–9, and the Raiders win, 26–20. What happens to your bet?

19. The New England Patriots are 3-point favorites over the New York Jets. The over-under line is 40. You make a $300 parlay bet on the Jets and "over."

    a. The Patriots win, 35–17. What happens to your bet?

    b. The Jets win, 17–7. What happens to your bet?

    c. The Patriots win, 28–27. What happens to your bet?

20. You make a $10 parlay card bet by picking 7 games. The payoff schedule on the card says "7 for 7 pays 80 for 1."

    a. Suppose all 7 of your picks cover the point spread on the parlay card. What happens to your bet?

    b. Suppose 6 of your 7 picks cover the point spread. What happens to your bet?

    c. Suppose P(each team covers) $= .5$. Find P(you win) and your expected winnings.

21. The Phoenix Cardinals are playing the New Orleans Saints. The money line is

Cardinals   +140
Saints   −160

a. You bet $160 on the Saints. The Saints win, 21–20. What happens to your bet?

b. You bet $160 on the Saints. The Cardinals win, 21–20. What happens to your bet?

c. You bet $100 on the Cardinals. The Saints win, 21–20. What happens to your bet?

d. You bet $100 on the Cardinals. The Cardinals win, 21–20. What happens to your bet?

22. The Chicago Bears are playing the Minnesota Vikings. The money line is:

Bears   +120
Vikings   −140

a. You bet $100 on the Bears. The Bears win, 34–28. What happens to your bet?

b. You bet $280 on the Vikings. The Vikings win, 21–20. What happens to your bet?

23. The Los Angeles Lakers are playing the Boston Celtics in a pro basketball game. The line is

Celtics
Lakers   6

The Lakers win, 108–104.

a. Which team is the favorite?

b. Which team is the home team?

c. What must happen for Lakers bettors to win?

d. You bet $22 on the Lakers. What happens to your bet?

e. You bet $33 on the Celtics. What happens to your bet?

24. The Golden State Warriors are playing the Orlando Magic.

Magic   220
Warriors   9½

The Warriors win, 134–117.

a. Which team is the underdog?

b. Which team is the favorite?

c. You bet $55 on the Warriors. What happens to your bet?

d. You bet $77 on the Magic. What happens to your bet?

e. You bet $110 on "over." What happens to your bet?

f. You bet $220 on "under." What happens to your bet?

25. The New York Knicks are playing the Houston Rockets.

Rockets   3
Knicks   205

The Knicks win, 103–102.

a. Which team is the underdog?

b. Which team is the favorite?

c. You bet $550 on the Rockets. What happens to your bet?

d. You bet $77 on the Knicks. What happens to your bet?

e. You bet $1100 on "over." What happens to your bet?

f. You bet $880 on "under." What happens to your bet?

26. In a baseball game between the Detroit Tigers and the New York Yankees, the money line is

> Tigers   +140
> Yankees   −160

a. You bet $160 on the Yankees. The Yankees win, 8–6. What happens to your bet?

b. You bet $160 on the Yankees. The Tigers win, 7–0. What happens to your bet?

c. You bet $100 on the Tigers. The Yankees win, 1–0. What happens to your bet?

d. You bet $100 on the Tigers. The Tigers win, 7–6, in 10 innings. What happens to your bet?

27. The Chicago Cubs are playing the New York Mets. The over-under line is:

> Cubs   8 over   +110
> Mets   8 under   −130

a. You bet $100 on "over." The Cubs win, 5–2. What happens to your bet?

b. You bet $130 on "under." The Cubs win, 5–2. What happens to your bet?

28. David is fighting Goliath. The odds are as follows:

> Goliath   −900
> David   +600

a. Suppose you bet $1800 on Goliath, and Goliath wins the fight. How much do you win?

b. Suppose you bet $1800 on David, and David wins the fight. How much do you win?

29. David is fighting Goliath. The odds that the fight will or will not go five rounds are as follows:

> Does go 5   −130
> Does not go 5   +110

a. You bet $260 that the fight will go five rounds. The fight ends in the fourth round when David knocks out Goliath with a shot to the head. What happens to your bet?

b. You bet $500 that the fight won't go 5 rounds. The fight ends 10 seconds into the fifth round. What happens to your bet?

30. There is a hockey game between the North Stars and the Canadiens. The betting line is as follows:

North Stars   +1½   +140
Canadiens   −1½   −160

  a. You bet $200 on the Stars. The Canadiens win, 6–5. What happens to your bet?

  b. You bet $240 on the Canadiens. The Canadiens win, 3–1. What happens to your bet?

  c. You bet $100 on the Stars. The Stars win, 6–5. What happens to your bet?

CHAPTER 6 • Blackjack

1. How many ways are there to pick 2 cards from 52? (*Hint*: Use the selection formula from the keno chapter.)

2. How many ways are there to pick 2 cards from 52 where one card is an ace and the other a 10-value card (ten, jack, queen, or king)?

3. Two cards are randomly dealt from an ordinary deck. What is the probability that they are a "natural"?

4. On the average, when playing blackjack, how often will you be dealt a natural? How often will the dealer be dealt a natural?

5. You bet $10. After all cards are played, the dealer's hand is ace, three, five. Your hand is nine, eight. What happens to your bet?

6. You bet $10. After all cards are played, the dealer's hand is eight, six, two, two, two. Your hand is six, nine, six. What happens to your bet?

7. You bet $20. You are dealt queen, ace. The dealer is dealt ten, four. What happens to your bet?

8. You bet $20. You are dealt seven, five. The dealer's up card is nine. You draw a jack. What happens to your bet?

9. You bet $40. You are dealt eight, three. The dealer's up card is six. You double down and are dealt a seven. The dealer's hole card is nine. The dealer draws a five. What happens to your bet?

10. You bet $60. You are dealt a pair of eights. The dealer's up card is seven. You split the pair. You are dealt nine on the first eight. You stand. You are dealt four on the second eight. You draw six. You stand. The dealer's hole card is a queen. What happens to your bets?

For Problems 11 through 18, follow the basic strategy.

11. You are dealt five, seven. The dealer's up card is nine. What should you do?

12. You are dealt ten, six. The dealer's up card is seven. What should you do?

13. You are dealt a pair of sixes. The dealer's up card is four. What should you do?

14. You are dealt six, four. The dealer's up card is seven. What should you do?

15. You are dealt seven, six. The dealer's up card is four. What should you do?

16. You are dealt ace, six. The dealer's up card is seven. What should you do?

17. You are dealt two, three. The dealer's up card is nine. You draw nine. What should you do?

18. You are dealt a pair of nines. The dealer's up card is seven. What should you do?

CHAPTER 7 • State Lotteries

1. Referring to the payoff table for the instant scratch-off game Win and Spin, find

   a. P(you win a $2 prize)

   b. P(you win a $5 prize)

   c. P(you win a $100 prize)

   d. P(you win a prize less than $500)

   e. P(you win no prize)

2. Here is the payoff table for the instant scratch-off game Wild Card:

Instant prize	No. of winners
$ 1	9,600,000
2	7,200,000
5	2,880,000
10	480,000
50	48,000
100	24,000
500	4627
1000	1200
5000	267
Go to Spin	50

   There were 120 million tickets printed. Find:

   a. P(you win a $100 prize)

   b. P(you win a $50 prize)

   c. P(you win a prize less than $50)

   d. Expected prize payoff (not including spin)

3. Suppose there is a Lotto game in which you pick 7 numbers from the numbers 1 through 50. Seven winning numbers are then selected at random. You win a prize if 4 or more of the numbers you picked are among the 7 winning numbers. Matching 7 for 7 is the jackpot. There is no bonus number. Find:

   a. P(you match 4 of 7)

   b. P(you match 5 of 7)

   c. P(you match 6 of 7)

   d. P(you win jackpot)

4. Suppose the Lottery Commission changes the game so that you must pick 13 numbers from 53 to win the jackpot. In this game, the chances of winning the jackpot are about 1 in a trillion. Suppose 20 million tickets are sold per game, with the usual two games per week. About how long will it take for somebody to win the jackpot?

5. One card is selected at random from an ordinary deck of cards. If the selected card is an ace, you win. Otherwise you lose. If you win, you are paid $10. What is the house edge if it costs you $1 to play the game?

6. Redo Problem 5, except, in addition, if the selected card is the queen of spades, you get a free replay.

7. Redo Problem 5, except in addition, if the selected card is any queen, you get a free replay.

8. Redo Problem 5, except in addition, if the selected card is anything but an ace or a king, you get a free replay.

9. Redo Problem 5, except in addition, if the selected card is anything but an ace, you get a free replay.

CHAPTER 8 • Horse Racing

1. In the fifth race at Upsand Downs, Brand New Bag comes in first, My Hero comes in second, and Fat Chance comes in third. After the race has been declared official, the tote board shows the following payoffs:

	Win	Place	Show
Brand New Bag	5.80	4.60	3.00
My Hero		7.20	4.40
Fat Chance			12.60

   a. How much do you win if you bet $2 on Brand New Bag to win? place? show?

   b. How much do you win if you bet $100 on My Hero to win? place? show?

   c. How much do you win if you bet $10 on Fat Chance to win? place? show?

2. The win pool for the seventh race at Upsand Downs is as follows:

Midnight Gambler	$ 18,000
Rush to Judgment	15,000
Lisa's Lament	2,000
Johnny Be Good	25,000
Mister Completely	19,000
Bettor Daze	28,000
Perry Mutuel	32,000
	$139,000

   a. What is the track take?

   b. Suppose Perry Mutuel wins the race. How much does a $2 win bet on Perry Mutuel pay?

   c. Suppose Bettor Daze wins the race. How much does a $2 win bet on Bettor Daze pay?

   d. Suppose Lisa's Lament wins the race. How much does a $2 win bet on Lisa's Lament pay?

   e. Suppose Lisa's Lament comes in second. How much does a $2 win bet on Lisa's Lament pay?

   f. Suppose Bettor Daze wins the race. How much does a $2 place bet on Bettor Daze pay?

g. Suppose Lisa's Lament comes in third. How much does a $2 win bet on Lisa's Lament pay?

3. The place pool for the seventh race at Upsand Downs is as follows:

Midnight Gambler	$ 19,000
Rush to Judgment	17,000
Lisa's Lament	2,500
Johnny Be Good	27,000
Mister Completely	20,000
Bettor Daze	25,000
Perry Mutuel	35,000
	$145,500

a. What is the track take?

b. Suppose Johnny Be Good wins the race and Rush to Judgment comes in second. How much do $2 place bets on Johnny Be Good and Rush to Judgment pay?

c. Suppose Rush to Judgment wins the race and Johnny Be Good comes in second. How much do $2 place bets on Johnny Be Good and Rush to Judgment pay?

d. Suppose Rush to Judgment wins the race and Lisa's Lament comes in second. How much do $2 place bets on Rush to Judgment and Lisa's Lament pay?

e. Suppose Rush to Judgment wins the race. How much do $2 place bets on Rush to Judgment pay?

4. The show pool for the seventh race at Upsand Downs is as follows:

Midnight Gambler	$ 23,000
Rush to Judgment	18,500
Lisa's Lament	3,500
Johnny Be Good	26,000
Mister Completely	22,500
Bettor Daze	27,000
Perry Mutuel	34,000
	$154,500

a. What is the track take?

b. Suppose Perry Mutuel wins the race, Johnny Be Good comes in second, and Mister Completely comes in third. How much do $2 show bets on each of these horses pay?

c. Suppose Bettor Daze wins the race, Lisa's Lament comes in second, and Johnny Be Good comes in third. How much do $2 show bets on each of these horses pay?

CHAPTER 9 • Prisoner's Dilemma

1. Here is the Prisoner's Dilemma payoff matrix:

$$
\begin{array}{cc}
 & \begin{array}{cc} \text{Cooperate} & \text{Be nasty} \end{array} \\
\begin{array}{c} \text{Cooperate} \\ \text{Be nasty} \end{array} &
\begin{bmatrix} (R, R) & (S, T) \\ (T, S) & (P, P) \end{bmatrix}
\end{array}
$$

where $T > R > P > S$, and $R > (T + S)/2$. Since $R > P$, why not cooperate in the one-move game?

2. Which of the following payoff matrices satisfy the conditions of Prisoner's Dilemma and which don't? Why?

a. $\begin{bmatrix} (2, 2) & (0, 3) \\ (3, 0) & (1, 1) \end{bmatrix}$ b. $\begin{bmatrix} (3, 3) & (0, 5) \\ (5, 0) & (4, 4) \end{bmatrix}$ c. $\begin{bmatrix} (2, 2) & (0, 5) \\ (5, 0) & (1, 1) \end{bmatrix}$

3. Suppose you play Prisoner's Dilemma for two moves, using the payoff matrix:

$$
\begin{array}{cc}
 & \begin{array}{cc} \text{Cooperate} & \text{Be nasty} \end{array} \\
\begin{array}{c} \text{Cooperate} \\ \text{Be nasty} \end{array} &
\begin{bmatrix} (3, 3) & (0, 5) \\ (5, 0) & (1, 1) \end{bmatrix}
\end{array}
$$

a. If your opponent uses Tit for Tat, which strategy gives you the highest score?

b. If your opponent is nasty on each move, regardless of what you do, which strategy gives you the highest score?

c. If your opponent cooperates on each move, regardless of what you do, which strategy gives you the highest score?

d. If your opponent tosses a coin on each move, cooperating if the coin lands heads and being nasty if the coin lands tails, which strategy gives you the highest score?

e. If your opponent cooperates on the first move, and is nasty on the second move, regardless of what you do, which strategy gives you the highest score?

f. If your opponent is nasty on the first move and cooperates on the second move, regardless of what you do, which strategy gives you the highest score?

g. Suppose your opponent is nasty on the first move, then cooperates on the second move if you cooperate on the first move, and is nasty on the second move if you are nasty on the first move. Which strategy gives you the highest score?

4. Consider a Prisoner's Dilemma game with three choices instead of two: cooperate, be nasty, and be really nasty. Suppose that the payoff matrix satisfies this general version of the Prisoner's Dilemma conditions:

(i) For any move you make, the nastier the opponent, the higher the opponent's score. Being really nasty is the dominant choice for both players.

(ii) No strategy pair has a higher payoff for both players than mutual cooperation.

(iii) Alternating noncooperative and cooperative choices with the opponent does worse than mutual cooperation.

Show that for the following three-move payoff matrix satisfying these generalized Prisoner's Dilemma conditions, a player does better against Tit for Tat by using the following strategy than by always cooperating:

Be nasty, really nasty, cooperative; nasty, really nasty, cooperative; nasty, really nasty, cooperative; . . .; etc.

	Cooperative	Nasty	Really nasty
Cooperative	(20, 20)	(10, 28)	(8, 30)
Nasty	(28, 10)	(18, 18)	(9, 26)
Really nasty	(30, 8)	(26, 9)	(12, 12)

ANSWERS TO FIGURING THE ODDS

CHAPTER 2 • Roulette

1. a. 8/38   d. $E = 5 \times -.158 = -.789$
    b. 30/38   e. House edge = 15.8%
    c. $E = -.158$

2. a. $E = 3 \times -.053 = -.159$
    b. House edge = 15.9/3 = 5.3%

3. a. 13/38   c. $E = -.289$
    b. 25/38   d. House edge = 14.5%

4.

	P(win)	P(lose)	Fair odds
a.	18/38	20/38	20 to 18 or 10 to 9
b.	2/38	36/38	36 to 2 or 18 to 1
c.	12/38	26/38	26 to 12 or 13 to 6
d.	4/38	34/38	34 to 4 or 17 to 2
e.	2/38	36/38	36 to 2 or 18 to 1

5. a. $E = 6 \times (-2/38) = -12/38$
    b. House edge = 5.3%

6. a. $E = 3 \times (-2/38) = -6/38$
    b. House edge = 5.3%

7. a.

Win	P(win)	
1	18/37	$E = -1/37 = -.027$
−1	19/37	House edge = 2.7%

   b.

Win	P(win)	
2	12/37	$E = -.027$
−1	25/37	House edge = 2.7%

   c.

Win	P(win)	
35	1/37	$E = -.027$
−1	36/37	House edge = 2.7%

   d.

Win	P(win)	
17	2/37	$E = -.027$
−1	35/37	House edge = 2.7%

8. a. $(20/38)^2 = .2770$
    b. $2 \times (18/38) \times (20/38) = .4986$

9. a. $(18/38)^3 = .1063$
    b. $(20/38)^3 = .1458$
    c. $3 \times (18/38)^2 \times (20/38) = .3543$
    d. $3 \times (18/38) \times (20/38)^2 = .3936$

10. a. $(18/38)^4 = .0503$
     b. $(20/38)^4 = .0767$
     c. $4 \times (18/38)^3 \times (20/38) = .2238$
     d. $6 \times (18/38)^2 \times (20/38)^2 = .3729$
     e. $4 \times (18/38) \times (20/38)^3 = .2762$
     f. $1 - P(\text{lose all 4 bets}) = 1 - (20/38)^4 = .9233$

11. a.

Win	P(win)	
1,000,000	1/10,000,000	$E = -0.15$
−.25	9,999,999/10,000,000	House edge $= .15/.25$
		$= 60\%$

     b.

Win	P(win)	
1,000,000	1/10,000,000	$E = -.40$
−.50	9,999,999/10,000,000	House edge $= .40/.50$
		$= 80\%$

12.

Win	P(win)	
1	21/38	
−1	17/38	$E = \$4/38 = \$.1053$

13.

Win	P(win)	
35	2/38	
−1	36/38	$E = \$34/38 = \$.8947$

14. a.

Win	P(win)	
38	1/38	
−1	37/38	$E = \$1/38 = \$.0263$

     b. Bet $1 on each of the 38 numbers. 37 will lose,
        but 1 will win. You will win $1 net.

15. a.

Win	P(win)	
20	0.99	$E = \$10$
−980	0.01	

     b. If the insurance company charges $10 per policy,
        its expected profits are zero; i.e., it breaks even.

CHAPTER 3 • Craps

1. a. $P(9) = 1/9$ d. $P(\text{odd}) = 1/2$
    b. $P(11) = 2/36$ e. $P(X > 7) = 15/36 = 5/12$
    c. $P(9 \text{ or } 11) = 1/6$ f. $P(X < 4) = 1/12$

2. a. $P(5) = 4/36$ e. $P(\text{even}) = 18/36$
       $P(7) = 6/36$ $P(\text{even before } 7) = 18/24 = 3/4$
       $P(5 \text{ before } 7) = 4/10$ f. $P(5,6,8,9) = 18/36$
    b. $P(6) = 5/36$ $P(7) = 6/36$
       $P(6 \text{ before } 7) = 5/11$ $P(5,6,8,9 \text{ before } 7) = 18/24 = 3/4$
    c. $P(2) = 1/36$ g. $P(8) = P(6) = 5/36$
       $P(2 \text{ before } 7) = 1/7$ $P(9) = P(5) = 4/36$
    d. $P(12) = 1/36$ $P(8 \text{ before } 9) = 5/9$
       $P(7 \text{ before } 12) = 6/7$ h. $P(\text{even before odd}) = 18/36 = 1/2$

3. a. $P(\text{win}) = P(4 \text{ or } 10) = 1/6$ d. House edge $= 16.7\%$
    b. $P(\text{lose}) = 5/6$ e. Fair odds $= 5$ to 1
    c. $E = -1/6 = -\$.167$

4. a. $P(\text{win}) = P(8,9,10)$
    $= 12/36 = 1/3$
    b. $P(\text{lose}) = 2/3$

    c. $E = 0$
    d. House edge $= 0\%$
    e. Fair odds $= 2$ to 1

5. a. $P(\text{win}) = P(6,7,8)$
    $= 16/36 = 4/9$
    b. $P(\text{lose}) = 5/9$

    c. $E = \$.222$
    d. $E = \$.111$
    e. Make the bet.

6. a. Win \$10.
    b. Win \$10.
    c. Lose \$10.
    d. Lose \$10.

    e. Win \$10.
    f. Lose \$10.
    g. Win \$10.
    h. No outcome.

    i. Lose \$10.
    j. Lose \$10.

7. a. Point is 6: $P(\text{win odds bet}) = P(6 \text{ before } 7) = 5/11$
    Fair odds $= 6$ to 5

    \$20 pass line bet wins \$20
       \$40 odds bet wins \$48

    Win \$68.

    b. Point is 9: $P(\text{win odds bet}) = P(9 \text{ before } 7) = 4/10$
    Fair odds $= 3$ to 2

    \$20 pass line bet wins \$20
       \$40 odds bet wins \$60

    Win \$80.

    c. Point is 10: $P(10 \text{ before } 7) = 1/3$
    Fair odds $= 2$ to 1

    \$20 pass line bet wins \$20
       \$40 odds bet wins \$80

    Win \$100.

    d. Lose \$60.
    e. Win \$20.

    f. Lose \$20.
    g. Win \$20.

    h. Lose \$20.
    i. Lose \$20.

8. a. Lose \$25.
    b. Lose \$25.
    c. Win \$25.

    d. Win \$25.
    e. Lose \$25.
    f. Lose \$25.

    g. Lose \$25.

9. a. Win \$30.
    b. Win \$30.
    c. Lose \$30.

    d. Lose \$30.
    e. Tie

    f. Win \$30.
    g. Win \$30.

10. a. Win \$40.
    b. $P(\text{win both}) = P(7) \times P(7) = (1/6)^2 = 1/36$
    c. $P(\text{win all } 3) = (1/6)^3 = 1/216$

11. a. Lose \$40.   b. Win \$80.   c. Win \$40.

12. a. Win \$36.   b. Lose \$20.   c. Lose \$20.

13. a. Lose \$30.   b. Lose \$30.   c. Win \$35.

14. a. Win \$70.   b. Lose \$60.   c. Win \$70.

15. a. Win \$60.   b. Lose \$60.   c. Win \$60.

16. a. Lose \$30.   b. Lose \$30.   c. Lose \$30.

17. a. Lose \$50.   b. Win \$350.   c. Lose \$50.

18. a, b, c, d. Lose \$70.

    e. Lose $75.
    f. Lose $335.
    g. Lose $555.
    h. Lose $455.
    i. Lose $555.
    j. The house edge for a combined $1 pass line bet and $1 odds bet whenever possible is 0.84%, so you lose $42.

19.   a. $P(\text{win}) = P(\text{win on come-out}) + P(\text{win by matching point})$
$$= 6/36 + P(\text{win by matching 2}) + \cdots + P(\text{win by matching 12})$$
$$P(\text{win by matching 2}) = P(\text{2 on come-out}) \times P(\text{2 before 7})$$
$$= (1/36) \times (1/7) = 1/252$$
$$= P(\text{win by matching 12})$$
$$P(\text{win by matching 3}) = P(\text{3 on come-out}) \times P(\text{3 before 7})$$
$$= (2/36) \times (2/8) = 1/72$$
$$= P(\text{win by matching 11})$$
$$P(\text{win by matching 4}) = 1/36$$
$$= P(\text{win by matching 10})$$
$$P(\text{win by matching 5}) = 2/45$$
$$= P(\text{win by matching 9})$$
$$P(\text{win by matching 6}) = 25/396$$
$$= P(\text{win by matching 8})$$
$$P(\text{win}) = 6/36 + 2(1/252 + 1/72 + 1/36 + 2/45 + 25/396)$$
$$= .473$$
   b. $P(\text{lose}) = 1 - .473 = .527$
   c. House edge $= 5.4\%$

20. $P(\text{2 before 7}) = P(\text{12 before 7}) = 1/7$
Fair odds $= 6$ to 1

$P(\text{3 before 7}) = P(\text{11 before 7}) = 1/4$
Fair odds $= 3$ to 1

21.   a. $P(\text{5 before 7}) = 4/10$ and fair odds $= 3$ to 2; win $95.
   b. Fair odds $= 3$ to 2; lose $65.
   c. Fair odds $= 6$ to 1; win $365.
   d. Fair odds $= 3$ to 1; lose $65.
   e. Win $365.
   f. Win $5.
   g. Win $185.

22.   a. $P(\text{you win } \$3) = (1/6)^3 = 1/216$
   b. $P(\text{you win } \$2) = 3 \times (1/6)^2 \times (5/6) = 15/216$
   c. $P(\text{you win } \$1) = 3 \times (1/6) \times (5/6)^2 = 75/216$
   d. $P(\text{you lose } \$1) = (5/6)^3 = 125/216$
   e. $E = -17/216$
   f. House edge $= 7.9\%$

23. *Game 1:*
$P(\text{you win}) = (5/6)^4 = .4823$
$E = -.0354$

    Game 2:
$P(\text{you win}) = (35/36)^{24} = .5086$
$E = .0172$

CHAPTER 4 • Keno and Slots

1. $280

2. $4

3. No payoff; you lose your $2 bet.

4. $45,000

5. House edge = 25%

6. a. $E = 2(.0839) + 20(.0121) + 1500(.0006) = 1.31$
    b. $E = -.69$
    c. House edge = 34.5%
    d. $P(\text{lose } \$2) = .9034$

7. $7! = 5040$

8. $11! = 39,916,800$

9. a. 3,628,800
    b. 120
    c. 120
    d. 120
    e. 1

10. $\dfrac{15!}{5!4!3!2!} = 37,837,800$

11. See Problem 10.

12. $\dfrac{11!}{7!4!} = 330$

13. $\dbinom{11}{7} = 330$

14. $\dbinom{11}{4} = \dbinom{11}{7} = 330$

15. $\dbinom{17}{9} = 24,310$

16. $\dbinom{17}{17} = 1$

17. $\dbinom{n}{k} = \dfrac{n!}{(n-k)!k!}$

18. $\dbinom{n}{n-k} = \dbinom{n}{k}$

19. $P(\text{3 of the selected panel members are gamblers}) = \dfrac{\dbinom{9}{3}\dbinom{21}{4}}{\dbinom{30}{7}}$

20. a. $\dfrac{\dbinom{13}{5}}{\dbinom{52}{5}}$

d. $\dfrac{\dbinom{13}{1}\dbinom{4}{3}\dbinom{12}{1}\dbinom{4}{2}}{\dbinom{52}{5}}$

b. $\dfrac{\dbinom{4}{1}\dbinom{13}{5}}{\dbinom{52}{5}}$

e. $\dfrac{\dbinom{13}{1}\dbinom{4}{3}\dbinom{12}{2}4^2}{\dbinom{52}{5}}$

c. $\dfrac{\dbinom{4}{3}\dbinom{4}{2}}{\dbinom{52}{5}}$

f. $\dfrac{\dbinom{13}{1}\dbinom{4}{2}\dbinom{12}{3}4^3}{\dbinom{52}{5}}$

21. a. $\dfrac{\dbinom{5}{3}\dbinom{15}{0}}{\dbinom{20}{3}}$

d. $\dfrac{\dbinom{5}{0}\dbinom{15}{3}}{\dbinom{20}{3}} = \dfrac{\dbinom{15}{3}}{\dbinom{20}{3}}$

b. $\dfrac{\dbinom{5}{2}\dbinom{15}{1}}{\dbinom{20}{3}}$

e. $1 - \dfrac{\dbinom{15}{3}}{\dbinom{20}{3}}$

c. $\dfrac{\dbinom{5}{1}\dbinom{15}{2}}{\dbinom{20}{3}}$

f. $E = 3 \times (1/4) = 3/4$

22. a. $\dfrac{\dbinom{7}{4}\dbinom{73}{16}}{\dbinom{80}{20}}$

c. $\dfrac{\dbinom{7}{6}\dbinom{73}{14}}{\dbinom{80}{20}}$

b. $\dfrac{\dbinom{7}{5}\dbinom{73}{15}}{\dbinom{80}{20}}$

d. $\dfrac{\dbinom{7}{7}\dbinom{73}{13}}{\dbinom{80}{20}}$

23. a. $\dfrac{\binom{12}{6}\binom{68}{14}}{\binom{80}{20}}$
   c. $\dfrac{\binom{12}{8}\binom{68}{12}}{\binom{80}{20}}$
   e. $\dfrac{\binom{12}{12}\binom{68}{8}}{\binom{80}{20}}$

   b. $\dfrac{\binom{12}{7}\binom{68}{13}}{\binom{80}{20}}$
   d. $\dfrac{\binom{12}{9}\binom{68}{11}}{\binom{80}{20}}$
   f. $\dfrac{\binom{12}{0}\binom{68}{20}}{\binom{80}{20}}$

24. a. $\dfrac{\binom{20}{0}\binom{60}{20}}{\binom{80}{20}}$
   c. $\dfrac{\binom{20}{10}\binom{60}{10}}{\binom{80}{20}}$
   e. $E = 20 \times (1/4) = 5$

   b. $\dfrac{\binom{20}{5}\binom{60}{15}}{\binom{80}{20}}$
   d. $\dfrac{\binom{20}{20}\binom{60}{0}}{\binom{80}{20}} = \dfrac{1}{\binom{80}{20}}$

25.

X	Reel 1 $P(X)$	Reel 2 $P(X)$	Reel 3 $P(X)$
Cherry	1/5	1/10	3/10
Grapefruit	3/10	1/5	1/10
Other	1/2	7/10	3/5

   a. $P(3 \text{ cherries}) = (1/5) \times (1/10) \times (3/10) = 3/500$
   b. $P(2 \text{ cherries, 1 grapefruit}) = [(3/10) \times (1/10) \times (3/10)]$
   $+[(1/5) \times (1/5) \times (3/10)]$
   $+[(1/5) \times (1/10) \times (1/10)] = 23/1000$
   c. $P(1 \text{ cherry, 2 grapefruit}) = [(1/5) \times (1/5) \times (1/10)]$
   $+[(1/10) \times (3/10) \times (1/10)] + [(3/10)$
   $\times (3/10) \times (1/5)] = 25/1000 = 1/40$
   d. $P(3 \text{ grapefruit}) = (3/10) \times (1/5) \times (1/10) = 3/500$

26. You lose about $90 per hour.

CHAPTER 5 • Sports Betting

1. a. Seahawks
   b. Seahawks
   c. Seahawks win by more than 6 points.
   d. Win $20.
   e. Lose $33.

2. a. Redskins
   b. Giants
   c. Win $50.
   d. Lose $77.
   e. Win $100.
   f. Lose $220.

3. a. Packers
   b. Broncos
   c. Lose $550.
   d. Win $70.
   e. Lose $1100.
   f. Win $800.

4. a. Dolphins   b. Lose $100.   c. Win $45.45.   d. Win $100.

5. a. Colts   b. Win $0 ($440 bet returned).   c, d, e. Win $0.

6.

	Falcons Rams 5 $110 on Falcons	Falcons Rams 3 $110 on Rams	Net, $
a.	Lose $110	Win $100	Lose $ 10
b.	Win 100	Win 100	Win 200
c.	Win 0	Win 100	Win 100
d.	Win 100	Win 0	Win 100
e.	Win 100	Lose 110	Lose 10

   f. Book gets middled.

7. E(winnings per $220 bet) $= .65(200) - .35(220) = \$53$
    E(winnings per $1 bet) $= \$.24$

8. E(winnings per $220 bet) $= .51(200) - .49(220) = -\$5.80$
    E(winnings per $1 bet) $= -\$.026$

9. a. $.045 per $1 bet
    b. $.045 per $1 bet
    c. No profit.

10. $[2.1(.60) - 1.1] \times 500 = \$80$

11. $[2.1(.58) - 1.1] \times 1000 = \$118$

12. $[2.1(.67) - 1.1] \times 560 = \$172$

13. $0.51 < 0.524$. Don't bet.

14. a.
Win	P(win)
5000	.55
−5500	.45

   $E = \$275$

   b. P(broke after up to 10 bets) $= 1 - (.55)^{10} = 0.997$

15. a. $E = .55(50) - .45(55) = \$2.75$ c. $E = 10(2.75) = \$27.50$
    b. 0   d. $E = 100(2.75) = \$275$

16. a. $E = \$110$ b. 0

17. a. Fraction of bankroll to bet $= 2.1(.55) - 1.1 = .055$ b. 0
       $E = \$15.13$

18. a. Win $130.   b. Lose $50.   c. Win $130.   d. Lose $50.

19. a. Lose $300.   b. Lose $300.   c. Win $780.

20. 80 for 1 is equivalent to 79 to 1 odds.
    a. Win $790.   b. Lose $10.

    c.
Win	P(win)
790	$.5^7 = 1/128$
−10	$127/128$

   $E = -\$3.75$

21. a. Win $100.   b. Lose $160.   c. Lose $100.   d. Win $140.

22. a. Win $120.   b. Win $200.

23. a. Lakers   d. Lose $22.
    b. Lakers   e. Win $30.
    c. Lakers must win by more than 6 points.

24. a. Magic   c. Win $50.   e. Win $100.
     b. Warriors   d. Lose $77.   f. Lose $220.

25. a. Knicks   c. Lose $550.   e, f. Bets returned.
     b. Rockets   d. Win $70.

26. a. Win $100.   b. Lose $160.   c. Lose $100.   d. Win $140.

27. a. Lose $100.   b. Win $100.

28. a. Win $200.   b. Win $10,800.

29. a. Lose $260.   b. Win $550.

30. a. Win $280.   b. Win $150.   c. Win $140.

CHAPTER 6 • Blackjack

1. $\binom{52}{2} = 1326$

2. $\binom{4}{1}\binom{16}{1} = 64$

3. $\dfrac{\binom{4}{1}\binom{16}{1}}{\binom{52}{2}} = \dfrac{64}{1326} = .048$

4. Both player and dealer will be dealt a natural about once every 21 hands.

5. Lose $10.

6. Win $10.

7. Win $30.

8. Lose $20.

9. Player bets $80 with hand totaling 18.
   Dealer has 20.
   Lose $80.

10. Player bets $120 with hands totaling 17 and 18.
    Dealer has 17.
    Win $60.

11. Draw.

12. Draw.

13. Split pair.

14. Double down.

15. Stand.

16. Draw.

17. Draw.

18. Stand.

CHAPTER 7 • State Lotteries

1. a. $P(\text{win } \$2) = .060$
    b. $P(\text{win } \$5) = .024$
    c. $P(\text{win } \$100) = .0002$

    d. $P(\text{win} < \$500) = .1686$
    e. $P(\text{win nothing}) = .83134$

2. a. $P(\text{win } \$100) = .0002$
    b. $P(\text{win } \$50) = .0004$

    c. $P(\text{win} < \$50) = .168$
    d. $E = .44$

3. a. $P(4 \text{ of } 7) = \dfrac{\binom{7}{4}\binom{43}{3}}{\binom{50}{7}} = .004324$

    b. $P(5 \text{ of } 7) = \dfrac{\binom{7}{5}\binom{43}{2}}{\binom{50}{7}} = .000190$

    c. $P(6 \text{ of } 7) = \dfrac{\binom{7}{6}\binom{43}{1}}{\binom{50}{7}} = .000003$

    d. $P(7 \text{ of } 7) = 1/99,884,400 = 0.00000001$

4. Each week 40 million lotto tickets are sold, so it takes about 25,000 weeks to sell 1 trillion tickets. Thus someone would hit the jackpot about once every 481 years.

5. House edge $= 23.1\%$

6. Think of the game as if it were being played with a deck of cards minus the queen of spades, i.e., played with a 51-card deck. We can look at it this way because when we draw the queen of spades, it sends us back into the game again. We can win or lose only if the final card drawn is anything but the queen of spades.

Result	Payoff	Win	P(result)
A	10	9	4/51
Other	0	−1	47/51

    House edge $= 21.6\%$

7. Think of the game as if it were being played with a full deck of cards minus all four queens, that is, played with a 48-card deck.

Result	Payoff	Win	P(result)
A	10	9	4/48
Other	0	−1	44/48

    House edge $= 16.7\%$

8. Think of the game as if it were being played with a full deck of cards minus all cards except aces and kings, that is, played with an eight-card deck consisting of only aces and kings.

Result	Payoff	Win	P(result)
A	10	9	1/2
Other	0	−1	1/2

House edge $= -400\%$

9. The game is being played with a four-card deck consisting of four aces.

In this game you win $9 with certainty.
House edge $= -900\%$

CHAPTER 8 • Horse Racing

1. *Note:* The amounts given represent profit; they do not include the amount
   of the bet.
   a. Win: Win $3.80.   b. Win: Lose $100.   c. Win: Lose $10.
      Place: Win $2.60.   Place: Win $260.   Place: Lose $10.
      Show: Win $1.00.   Show: Win $120.   Show: Win $53.

2. a. $23,630
   b. Win $7.20.
   c. Win $8.20.
   d. Win $115.20.
   e. Lose $2.
   f. Not given place pool
      in this problem.
   g. Lose $2.

3. a. $24,735
   b. Johnny Be Good: $4.80.
      Rush to Judgment: $6.40.
   c. Place bets do not depend on the order of the two horses that place. The
      answer is the same as in part (b).
   d. Rush to Judgment: $7.80.
      Lisa's Lament: $42.40.
   e. Cannot be answered without learning the 2nd place finisher.

4. a. $26,265
   b. Perry Mutuel: $2.80.
      Johnny Be Good: $3.00
      Mister Completely: $3.20.
   c. Bettor Daze: $3.60.
      Lisa's Lament: $15.60.
      Johnny Be Good: $3.80.

CHAPTER 9 • Prisoner's Dilemma

1. No matter what the opponent does, it's better to be nasty than to cooperate
   (T > R, P > S). So if you have only one encounter, be nasty.

2. a. 3 > 2 > 1 > 0 and 2 > (3 + 0)/2
      Conditions of Prisoner's Dilemma are satisfied.
   b. P = 4 > 3 = R
      Conditions of Prisoner's Dilemma are *not* satisfied.
   c. 5 > 2 > 1 > 0 but 2 < (5 + 0)/2
      Conditions of Prisoner's Dilemma are *not* satisfied.

3.

	1st move	2nd move
a. Tit for Tat	C	C
You	C	N
Score = 8		

	1st move	2nd move
b. All N	N	N
You	N	N
Score = 2		

	1st move	2nd move
c. All C	C	C
You	N	N
Score = 10		

d. The best strategy is to be nasty on both moves, maximizing expected payoff.

e. Nasty, nasty. Score = 6

f. Nasty, nasty. Score = 6

g.

	1st move	2nd move
Opponent	N	C
You	C	N
Score = 5		

4. Consider the first three moves. Always cooperating against Tit for Tat gives a player the scores 20, 20, 20 (60 points). Being nasty, really nasty, cooperative against Tit for Tat gives scores 28, 26, 8 (62 points). This three-move cycle keeps repeating. Thus whenever the game stops, the cycle "nasty, really nasty, cooperative, . . ." yields a higher total score against Tit for Tat than always cooperating.

INDEX

Averages, law of, 3
Axelrod, Robert, 136, 138

Baseball, 84
Basketball, 82
Betting layouts (casino)
  blackjack, 88
  craps, 28
  roulette, 8
Blackjack, 87–96
  basic strategy, 91–94
  betting layout, 88
  card counting, 91, 94–95
  money management, 94
  multiple deck games, 87, 91
  rules of play, 87–91
Bookie, 55
Boxing, 82–83
Breakage, 117
Breiman, Leo, 77

California Lottery, 97–111
  annuity system, 105
  Lotto, 100–109
    Decco, 109
    the new California Lotto, 107
    payoff odds, 103–105
    probabilities, 103–105, 108–111
    Topper, 108
  Scratch-off, 97–100
    Big Spin, 97, 99–100
    Win and Spin, 98
  systems, 105–107

Casino betting layouts.
  See Betting layouts
Churn, 13
Combinatorics, 48–50
Craps, 27–46
  any craps bet, 40
  bet on 7, 39
  bet on 2, 3, 11, or 12, 39–40
  betting layout, 28, 36, 37
  big 8 bet, 43
  big 6 bet, 43
  buy bet, 45
  come bet, 33–34
  come-out roll, 30, 33–34
  craps-eleven bet, 41
  don't come bet, 35–36
  don't pass bet, 34–35
  field bet, 41
  hardway bet, 44
  horn bet, 41
  house edge, 46
  lay bet, 45
  odds bet, 36–38
  one roll bet, 39–41
  pass line bet, 30–33
  place bet, 42

DeMère, Chevalier, 2
Dice combinations, 29
Double-up strategy, 21–24

Even odds, 10
Expected winnings
  computation of, 11–13
  probability distribu-tion table, 11

Fair odds, 13, 14
Fixed-fraction betting, 77–78
Football, 56–80
  futures bet, 71, 72
  middles, 60
  Monday night home underdogs, 74–75
  money line bet, 69–71
  moving the spread, 59
  opening line, 58
  over-under bet, 61
  parlay bet, 62
  parlay cards, 63–66
  point spread, 56–60
  sampling, 74
  splitting the action, 58
  teaser cards, 66–68
  weird bets, 71, 73
Fuller, Lisa, 179

Gombaud, Antoine, 2

Hockey, 85
Horse racing, 112–129
  breakage, 117
  claiming races, 121
  Daily Racing Form, 122–124
  exotic bets, 120
  maiden races, 121
  minus pools, 120
  off-track betting, 125
  overlays, 125
  pari-mutuel system, 113
  place bet, 117–119
  program, 125–127
  show bet, 119–120
  stakes races, 121–122
  steam, 128

Horse racing (*continued*)
   track take, 113
   win bet, 116–117
House edge
   craps, 46. *See also*
      specific types of
      bets
   definition, 4
   football bets, 57
   keno, 48, 53
   law of averages, and,
      10
   point spread bet, 57
   roulette, 10, 19, 25
   slots, 54

Independent events, 3,
   23

Kelly, J. L., Jr., 77
Kelly system, 76–78, 94
   fixed-fraction betting,
      77–78
Keno, 47–53
   house edge, 48, 53
   probabilities, 51–53
   10-number bet, 47

Las Vegas Sports
      Consultants, 58
Law of averages, 3
Lollapalooza, the, 95
Lotteries, state. *See*
      California Lottery

Middles, 60
Money line bet, 67, 70,
   82–85
Money management,
   76–78, 94
   fixed-fraction betting,
      77–78

Odds
   even, 10
   fair, 13–14, 36–39,
      67

giving odds, 67
payoff, 9
taking odds, 67
Opening line, 58

Parlay bet, 62
Parlay cards, 63–66
Payoff odds, 9
Permutations and
      combinations,
      48–50
Place bet
   craps, 42
   horse racing, 114,
      117–119
Point spread, 56–60
   basketball, 80
   middles, 60
   money line bet, 70, 71
   moving the spread, 59
   opening line, 58
Predictability, 74–75
Prisoner's Dilemma,
      131–139
   computer tourna-
      ments, 136
   general version, 132
   iterated, 133
   repeated play,
      133–136
   strategies, 134–139
   survival of the fittest,
      138
   Tit for Tat, 135,
      136–139

Racing horse. *See* Horse
   racing
Rapoport, Anatol, 136
Roulette, 7–25
   bets on numbers
      1–18 or 19–36,
      19
   betting layout, 8
   churn, 13
   column bet, 18
   combination bet,
      19–20

double-up strategy,
   21–24
dozens bet, 17–18
even and odd bet, 18
five numbers bet,
   16–17
four numbers bet, 16
house edge, 10, 19, 25
red and black bets,
   11–12, 18
single numbers bet,
   14–15
six numbers bet, 17
three numbers bet, 15
two numbers bet, 15
Roxborough, Michael
      ("Roxy"), 1, 58,
      66, 76

Sampling, 74
Scams
   the Big Player
      concept, 94
   the card counting
      computer, 95
   the Lollapalooza, 95
   on-track illegal
      betting, 129
   steam, 128
Slots, 53–54
   house edge, 54
   probabilities, 54
Sports betting, 55–86
   baseball, 84
   basketball, 82
   bookies, 55
   boxing, 82–83
   football (*see* Football)
   hockey, 85
   Kelly system, 76–78,
      94
   media coverage,
      80–82
   money management,
      76–78
   opening line, 58
   predictability, 74

splitting the action, 58
sports books, 55
sports services, 79
Sports services, 79
State lotteries. *See*
    California Lottery
Strategies and systems
    basic strategy
        (blackjack),
        91–94

California Lottery,
    105–107
double-up, 21–24
Kelly system, 76–78,
    94
middles, 60
Prisoner's Dilemma,
    134–139
roulette, double-up,
    21–24

Thorp, Edward O., 87,
    92
Track take, 112
Tucker, A. W., 130
"21." *See* Blackjack

Video poker, 54
Vigorish, 58